# The patent system
# and inventive activity
### during the
### industrial revolution
### 1750–1852

A caricature of the original Patent Office Library known as the 'drain-pipe'. From H. Harding *Patent Office Centenary* (London, 1953)

# H. I. DUTTON

# The patent system and inventive activity

## during the industrial revolution 1750–1852

*Manchester University Press*

Copyright © Harold Irvin Dutton 1984

Published by
Manchester University Press
Oxford Road, Manchester M13 9PL, U.K.
51 Washington Street, Dover, N.H. 03820, U.S.A.

*British Library cataloguing in publication data*
Dutton, H. I.
   The patent system and inventive activity during
   the industrial revolution 1750–1852.
   1. Patent laws and legislation—Great Britain—History
   2. Great Britain—Industries—History
   I. Title
   338′.0941      HC253
   ISBN 0-7190-0997-9

*Library of Congress Cataloging in Publication Data*
Dutton, H. I.
   The patent system and inventive activity during
   the Industrial Revolution, 1750–1852.
   Bibliography: p.
   Includes index.
   1. Patents—Great Britain—History.
   2. Inventions—Great Britain—History.
   I. Title. II. Title: Industrial Revolution.
   T257.P2D87   1984      608.741      83-18803
   ISBN 0-7190-0997-9

Printed in Great Britain
by Butler & Tanner Ltd, Frome and London

# Contents

|   |   |   |
|---|---|---|
| List of tables | *page* | vi |
| Abbreviations | | vi |
| Acknowledgements | | vii |
| Introduction: Problems and sources | | 1 |

## Part I  The patent institution

| | | |
|---|---|---|
| 1. | The case for the patent system | 17 |
| 2. | The objectives of patent reform and the emergence of the invention interest | 34 |
| 3. | The Patent Law Amendment Act, 1852 | 57 |
| 4. | Patent law and the courts | 69 |
| 5. | Patent agents: the early growth of a nineteenth-century service sector | 86 |

## Part II  Patents and inventive activity

| | | |
|---|---|---|
| 6. | Invention and inventive activity | 103 |
| 7. | Trade in invention | 122 |
| 8. | Investment in patents | 150 |
| 9. | Patentees, competition and the law | 175 |
| | Conclusion | 202 |
| | Appendices | 206 |
| | Bibliography | 211 |
| | Index | 225 |

# List of tables

| | | |
|---|---|---|
| 1. | English patents sealed, 1750–1851 | *page* 2 |
| 2. | The number of patent law cases, 1770–1849 | 71 |
| 3. | Percentage of reported cases going for and against patentees at common law and at Equity, 1750–1849 | 78 |
| 4. | The number of patents taken out in Manchester and Birmingham, 1723–1852 | 89 |
| 5. | Percentage of total patents sealed, 1751–1850, taken out by multiple patentees | 114 |
| 6. | The industrial spread of multiple patentees, 1751–1852 | 115 |
| 7. | The industrial spread of multiple patentees holding four or more patents, 1751–1852 | 116 |
| 8. | Companies working patents, 1837–52 | 164 |
| 9. | Annual turning-points in patents and the British trade cycle, 1798–1850 | 177 |
| 10. | Output of yellow paint, 1783–90 | 179 |

# Abbreviations

Add. MSS.   British Museum Additional Manuscripts
B.M.   British Museum
J.H.C.   Journal of the House of Commons
L.J.   Lords Journal
P.R.O.   Public Record Office

# Acknowledgements

The preparation of this study has involved a long journey, and on the way I have been helped by a number of people. First of all I would like to thank Professor D. C. Coleman, who encouraged me to pursue this unexplored subject when I was a graduate student. I have also benefited from the advice of Professor T. C. Barker, Professor W. H. Chaloner, Professor W. R. Cornish, the late Professor A. H. John and Oliver Westall, all of whom read different versions and parts of this work. I am grateful to the many archivists and librarians who helped me as I visited one Record Office after another. My thanks also to Sir Charles Carter and to London University for financing research expenses, and to Paul McGloin for his help. The friendly Economics Department at the University of Auckland not only made my year in New Zealand a delightful experience but also allowed me time to write the book — and to fish for the big trout of the magnificent Tongariro! I would like to thank my colleagues at the University of Lancaster, especially John Channon and Mary Rose for reading the proofs, and to acknowledge the debt I owe John King. Finally, I would like to thank my friend Dr Stephen Jones for his help and guidance over the past ten years. Many of the ideas developed in chapter 6 have grown out of our joint research on the British pin industry, which is still in progress. Needless to say, I am responsible for all errors of fact and interpretation.

For Hilly

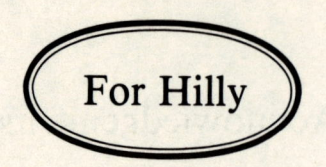

# Introduction: Problems and sources

The operation of the patent system during the industrial revolution has rarely been the subject of scholarly research.[1] Some economic historians have examined particular aspects of the system and how it affected individual inventors, and some have used patent statistics, but there has been no general study of the patent system as a whole between 1750 and 1850.[2] In 1946 Gomme published a brief account of the origins and growth of the patent system in Britain, and in 1953 Harding celebrated the centenary of the 1852 Patent Law Reform Act with a short history of the system, but, whilst both are useful, they concentrate almost exclusively on administrative changes rather than on the kind of questions which would interest economic historians.[3] 'Little attempt,' Harding observed, 'appears to have been made to assess the part played by patents in the development of the industries of Great Britain,'[4] and this remains the case.

The lack of research on the patent system during the industrial revolution is all the more puzzling considering the steady increase in the number of patents. In 1750 only seven were registered. By 1851 the annual number had increased to 455. In between, 13,227 patents were taken out to protect inventions in England,[5] and significantly this activity varied between industries and within industries over time, a matter which certainly calls for some explanation. The vast general literature on patents simply adds to the puzzle. Patents in pre-industrial England have attracted much attention, largely because of their association with the political problems during Elizabeth's reign, and the modern patent system has been even more thoroughly examined by Boehm, Silberston and Taylor.[6] In the middle, though, there is a wide uncultivated space. The historiographical emphasis on the importance of invention and innovation deepens the puzzle still further. From Samuel Smiles, who saw the industrial revolution almost entirely in terms of the heroic inventor, right through to the late Ralph Davis's book on foreign trade, inventive activity has, in

## Introduction

*Table 1   English patents sealed, 1750–1851*

| Date | Patents | Date | Patents | Date | Patents | Date | Patents | Date | Patents |
|---|---|---|---|---|---|---|---|---|---|
| 1750 | 7 | 1771 | 22 | 1792 | 85 | 1813 | 131 | 1834 | 207 |
| 1751 | 8 | 1772 | 29 | 1793 | 43 | 1814 | 96 | 1835 | 231 |
| 1752 | 7 | 1773 | 29 | 1794 | 55 | 1815 | 102 | 1836 | 296 |
| 1753 | 13 | 1774 | 35 | 1795 | 51 | 1816 | 118 | 1837 | 256 |
| 1754 | 9 | 1775 | 20 | 1796 | 75 | 1817 | 103 | 1838 | 394 |
| 1755 | 12 | 1776 | 29 | 1797 | 54 | 1818 | 132 | 1839 | 411 |
| 1756 | 3 | 1777 | 33 | 1798 | 77 | 1819 | 101 | 1840 | 440 |
| 1757 | 9 | 1778 | 30 | 1799 | 82 | 1820 | 97 | 1841 | 440 |
| 1758 | 14 | 1779 | 37 | 1800 | 96 | 1821 | 109 | 1842 | 371 |
| 1759 | 10 | 1780 | 33 | 1801 | 104 | 1822 | 113 | 1843 | 420 |
| 1760 | 14 | 1781 | 34 | 1802 | 107 | 1823 | 138 | 1844 | 450 |
| 1761 | 9 | 1782 | 39 | 1803 | 73 | 1824 | 180 | 1845 | 572 |
| 1762 | 17 | 1783 | 64 | 1804 | 60 | 1825 | 250 | 1846 | 493 |
| 1763 | 20 | 1784 | 46 | 1805 | 95 | 1826 | 141 | 1847 | 493 |
| 1764 | 18 | 1785 | 61 | 1806 | 99 | 1827 | 150 | 1848 | 388 |
| 1765 | 14 | 1786 | 60 | 1807 | 94 | 1828 | 154 | 1849 | 514 |
| 1766 | 31 | 1787 | 55 | 1808 | 95 | 1829 | 130 | 1850 | 513 |
| 1767 | 23 | 1788 | 42 | 1809 | 101 | 1830 | 180 | 1851 | 455 |
| 1768 | 23 | 1789 | 43 | 1810 | 108 | 1831 | 151 | | |
| 1769 | 36 | 1790 | 68 | 1811 | 115 | 1832 | 147 | | |
| 1770 | 30 | 1791 | 57 | 1812 | 118 | 1833 | 180 | | |

*Source.* B. R. Mitchell and P. Deane, *Abstract of British Historical Statistics,* Cambridge, 1962, pp. 268–69.

varying degrees, frequently been cited as one of the more significant explanations of the industrial revolution.[7] Yet very few have attempted to analyse the institutional and legal framework within which this inventive activity took place, despite the fact that institutional change and property right theory are now the rage. In the survey of studies of technology in economic history by Uselding there is no mention of patents, or inventive activity.[8] Patents, as one recent writer has correctly observed, have been *terra incognita* to many economists and economic historians, especially on this side of the Atlantic.[9]

This lack of research has not prevented many from passing judgements. In fact, as Eric Robinson notes, 'there is hardly one [historian] who has not felt it incumbent to express some opinion about the matter'.[10] Boehm and Silberston, for example, have argued that the industrial revolution was the 'age of patentless invention', and that patents were largely irrelevant as a means of inducing inventions. They claim, consequently, that some other stimuli 'must have been responsible for the inducement of a large body of nineteenth century invention'.[11] T. S. Ashton is rather more tentative on this question, but concludes in much the same way: 'It is at least possible that without the apparatus of the patent system discovery might have developed quite as rapidly as it did.'[12] Landes, who confines his comments to a footnote, doubts whether patents were significant during the industrial revolution, and cites a number of reasons: this kind of protection was not new; the cost and difficulty of obtaining patents increased steadily; patents were too easily made ineffective by determined competitors; and entrepreneurs relied upon secrecy rather than on the protection provided by the law.[13]

Others take a different view. Holdsworth, for example, claims that during the late seventeenth and eighteenth centuries 'the administration of the law as to the grant of patents ... was successful in encouraging British industry'.[14] Fox considered that 'it was ... not by accident that the patent system had its origins in England nor that the Industrial Revolution was the inevitable consequence'. The patent system, he concludes, 'has been one of the greatest of all elements which have contributed towards the expansion of industry and the development of science and the useful arts'.[15] In a much neglected book Ravenshear argues that 'patents exercised a net influence in stimulating the growth of industry'.[16] For Harding there was 'little doubt that patents helped to create the industrial supremacy which existed at the time of the Great Exhibition'.[17] And Hatfield, writing in a mood of patriotic zeal, concluded that the 'patent law was our invention, and it gave us the first place among nations in industry for over 200 years'.[18]

Nor do these different arguments arise simply because of the lack of research. Patents and the patent system also pose a number of intractable conceptual and empirical problems. For a start, economic theory is not very helpful in assessing the net economic effects of this institution. In fact Joan Robinson goes so far as to say that the 'patent system introduces some of the greatest complexities in the capitalist rules of the game because it is rooted in a contradiction'.[19] Simply stated, the contradiction arises for the following reason. Patents are worth having because they encourage inventive activity and thus, ultimately, the diffusion of best-practice techniques. But to ensure that inventive activity increases, the rate of diffusion has to be artificially slowed down to allow inventors time to appropriate sufficient returns to make the investment in invention worthwhile. From society's point of view, a high rate of inventive activity *and* a high rate of diffusion would be the ideal. But it is not possible to have both simultaneously, because knowledge — which by definition is embodied in inventions — is privately produced, and if it becomes public property at the moment it is made, the rate of return for the producer will be too small. Knowledge, and consequently invention, are a public good. Hence it is not used up in the process of being utilised, but any restrictions on the use of existing knowledge lead to a less than optimal use and allocation of resources. Consequently, a freely competitive market does not provide sufficient incentives to increase the supply of knowledge and invention, whilst an imperfect market, where restrictions are imposed, leads to underutilisation and inefficiencies. At one extreme, the absence of patent protection in the form of a temporary monopoly may mean there are no inventions to diffuse. This is an exceedingly unlikely outcome, but there is no doubt that patents and Pareto conditions for optimality do not mix.[20]

Theoretically, socially optimal patent protection is achieved when the marginal social costs of patent protection are equal to its marginal social benefits. Yet these costs and benefits are not always obvious and can change over time. According to Rosenberg, the

> benefits of the patent system are the increase in output which society owes to that class of inventions which would not have been made at all in the absence of the system, plus the increase in output owing to the earlier introduction of inventions which would have come anyway, but at a later date. The costs of the system are the restrictions in output for which the patent laws are responsible by allowing all holders of patent inventions to exclude others from their use during the period of patent protection.[21]

Although he readily admits the impossibility of establishing 'the precise magnitude of these costs and benefits', this assessment omits a number of different problems. Three examples will be sufficient to demonstrate the difficulties. Firstly, increases in output may simultaneously increase the level of pollution, and this negative externality may also increase in a non-linear way with output. Patents, therefore, may encourage overinvestment in inventions in areas which are socially undesirable, and any cost-benefit analysis would have to quantify the *real* as opposed to the nominal level of output.[22] A similar and second problem was identified by Sir Arnold Plant when he argued that patents may divert inventive activity into areas which are patentable rather than those which may be socially desirable, but essentially non-patentable.[23] Finally, it is quite possible that patents, whilst restricting the use of knowledge, may in fact increase the level of competition by allowing producers to rival *existing* monopolists. In this sense they may, paradoxically, act as an anti-monopoly device.[24] All this is not meant as a criticism of Rosenberg, who has made a pioneering contribution to both the theory and the history of technological change, but to show the immense conceptual problems and paradoxes which any analysis of the patent system presents.

The recent work on the optimal life of a patent, which has been seen as one way of bringing social and private costs and benefits into line without abolishing the patent system altogether, has not proved successful in practical policy terms.[25] Even though there is, theoretically, a socially optimum length of patent protection, it is impossible to determine this optimum *ex ante*, unless the significance of an invention is known at the time it is made.[26] As technology becomes more interdependent it would also require an assessment of the significance of other inventions, since one invention may change in significance as others are produced. In short, a global analysis of the patent system is subject to all manner of conceptual difficulties which effectively preclude anything but intelligent guesses. As far as the costs and benefits of the patent system are concerned, Machlup concludes thus: 'No economist, on the basis of present knowledge, could possibly state with any certainty that the patent system, as it now operates, confers a net benefit or a net loss upon society. The best he can do is to state assumptions and make guesses about the extent to which reality corresponds to these assumptions.'[27]

If it were possible to specify a model of the British economy without a patent system many of these problems would disappear, but to do so would involve stretching counterfactual history to impossible extremes. A comparison with a country without a patent system could provide a second type of solution, but all the important industrial countries in the nineteenth century adopted, some with significant variations, the basic features of the British system. Had the mid-nineteenth-century movement to abolish the patent system succeeded, then perhaps another story could have been written.

Schiff's study of inventive activity in Switzerland and the Netherlands, which abolished their patent systems in 1850 and 1869 respectively, indicates what could have been done.[28] His study is based on what he calls a 'differential historical analysis'. He compares industrial development in Switzerland and the Netherlands with simultaneous industrial development in patent-granting countries. He also compares industrial development in both these countries with and without a patent system. His study is concerned with answering two important questions. Firstly, do patents appreciably increase the level of inventive activity? Secondly, if patents do increase the volume of invention, does this increase the speed of technological change and the pace of industrial development? His findings regarding the first question are inconclusive. The reintroduction of the patent system in the Netherlands in 1912 supports the view that patents do act as an incentive to inventive activity. In the case of Switzerland, where the patent system was reintroduced by stages in 1888 and 1907, it seems that the volume of inventive activity was not appreciably different with or without the patent system. On the second question he finds that during the absence of the patent the pace of industrial development in the Netherlands was not appreciably slower, and that in Switzerland industrial development was rather more vigorous. He concludes his study with the argument that the 'industrialisation of a country can proceed smoothly and vigorously without a national patent system'. This is not the place to offer a critique of Schiff's book, but his conclusion, though interesting, is not very profound, and it may well be that both economies could have grown faster with a patent system.[29]

Apart from these conceptual difficulties, there are also serious problems with the historical evidence and data. Some are general in nature and some are unique to the period of the industrial revolution. The most serious problem is that patents are not synonymous with inventive activity or invention. Every writer on the subject has experienced the despair which this causes. Some have simply rejected them as a useful source, others have used them cautiously, and some have accepted that they are virtually the only means of analysing inventive activity. There are a number of alleged deficiencies which need to be noted. The most serious is that patent statistics do not reflect the *quantity* of inventive output, because for a number of reasons (which will be considered later) inventors prefer to keep their inventions secret rather than pay for the (allegedly) expensive protection of a patent. A few, during the early years of the industrial revolution, may also have been ignorant of how to acquire patent protection.[30] Patent statistics, moreover, do not reflect the *quality* of inventive output, because they treat the considerably varied nature and value of inventions equally. This, though, is less serious than is supposed: the problem of quality comparisons exists independently of patents.[31] On the other hand it causes a problem if patents are taken out for inventions which in fact are not inventions at all. This depends, of course, on the standards of patentability employed by the Patent Office officials. If standards are low, as was the case between

# Introduction 7

1750 and 1852 (and after), it is quite possible that the number of patents *overstates* the level of inventive activity.[32] Despite these limitations (which will be discussed in more detail later), patent data are the only index available to indicate the volume of inventive activity and should not be abandoned because they are imperfect. Schmookler makes the point simply: 'We have a choice of using patent statistics cautiously and learning what we can from them, or not using them and learning nothing about what they alone can teach us.'[33]

In addition to the problems associated with the use of patent statistics, the evidence on which this study is based is limited in a number of ways. Few inventors left records explaining why they invented and why they invented in certain areas. Most of the manuscript material found in business deposits is concerned largely with legal technicalities and with the nuts and bolts of technology itself, rarely with questions that would interest the economic historian, such as motives, costs and rewards. Other contemporary sources are in many ways equally unhelpful. The autobiographies by Bessemer, Nasmyth and Bramah are exceptional, but most 'lives' of inventors, a popular form of literature in the nineteenth century, are uneven in quality and all too often pandered to the Victorians' pious sense of progress. The mechanical journals and technical encyclopaedias which burgeoned in the first half of the nineteenth century contain a good deal of information, but it is mostly technical in nature, and many simply reprinted full or abridged versions of the specifications enrolled in the Patent Offices. Significantly, biographical material is generally left for the obituary, bankruptcy and letter columns. The three Select Committees on Patents in 1829, 1835 (which met behind closed doors and never published its findings) and 1851 provide a great deal of material on attitudes and arguments concerning the relationship between patents and inventive activity, but few hard facts. Patent court cases are just as infuriating to use, if only because they always seem more promising than they actually are.

In short, useful evidence is thin and patchy, which makes it all the more difficult to assess what is typical. It also precludes an examination of some important questions concerning the relationship between market structure, inventive activity, patents and technological advance. Although recent work in industrial economics has shown that Schumpeter's argument, in which technological change requires the existence of large firms with some transient degree of market power, is inconclusive on *a priori* grounds, it would have been an interesting exercise to test this hypothesis for the period of the industrial revolution. Unfortunately the data are inadequate for the purpose, and so questions concerning the role of patents in controlling the growth and development of certain industries remains unanswered — for the present at least.[34] The shortcomings of the existing evidence have, as a result, shaped the nature of this study rather more than was originally hoped. There is, of course, nothing especially unique about that: records were hardly ever created for posterity, let alone for historians learning a difficult craft. The findings of this study do, however, lend support to Jewkes's perceptive comment that the 'study of invention must necessarily be qualitative, selective, impressionistic, more historical than scientific'.[35]

The primary aim of the present study is to assess the impact of the patent system on inventive activity. Very little will be said about innovation, except in a rather peripheral way. This is not to suggest that innovation is unimportant. In fact, this aspect of technological change (together with the factors which determine the rate of diffusion) has been the major focus of the literature ever since the publication of Habakkuk's *American and British Technology in the Nineteenth Century* in 1962. If this makes the present study rather lopsided, there is ample reason for believing that the lack of research on patents and inventive activity has made studies on innovation equally lopsided. Technological change is an extremely complex process, and if it is to be understood analysis should perhaps proceed by small, narrowly defined steps. This is especially the case when considering a long time period and when the economy is itself experiencing structural change.

This deliberately narrow focus is based on the notion that technological change is the result of a number of separate processes which are conceptually quite distinct. Theoretically it is possible to distinguish between a number of key sequential propensities: the propensity to develop pure science, to invent, to innovate, to finance innovation and, finally, the propensity to accept innovation.[36] Empirically there are some difficulties with these distinctions, especially if all the processes are carried out by an individual, or by a single firm. Here, inputs and outputs in the process of technological change are less easily discernible, at least for the purpose of measurement. As will be shown later, it is often impossible to distinguish, for example, between the finance of invention and the finance of innovation. There are also conceptual problems. In Usher's 'cumulative synthesis approach', where invention requires an 'act of insight' beyond the normal exercise of technical and professional skill, there is no clear distinction between the processes of invention and innovation. In fact, in the final stage of 'critical revision' where all perceived relations between separate elements become fully understood, invention and innovation are one and the same thing.[37] There are others who see little advantage in separating invention from innovation,[38] but, historically at least, there is merit in making the distinction because the two processes were often carried out by different people.[39]

The definition of invention adopted for this study is provided by Jacob Schmookler (whose work will be discussed in Chapter 6). He defines invention as a new way of producing something old (that is, a process invention), or an old way of producing something new (that is, a product invention). Every invention, therefore, 'is a new combination of pre-existing knowledge which satisfies some want'. Although in practice some inventions alter the process of production as well as changing the product in a qualitative way, this definition covers all the possibilities. Innovation, on the other hand, is defined here as the *first* use of an invention, whilst diffusion refers to the rate at which the innovation is imitated by other users.[40]

This study is in two parts. The first will examine the patent institution itself. Chapter 1 will discuss contemporary attitudes towards the patent system and the arguments used to justify its continued existence. Chapters 2 and 3 will examine the development of the patent reform movement and the arguments which were put forward to alter the way in which the patent system worked. Chapter 4 will assess the way in which judges decided on patents when they were contested in the courts, and the final chapter in this section discusses the early growth of the profession of patent agent. These matters are treated in some depth for two reasons. Firstly, to understand the institutional framework within which inventive activity took place it is necessary to examine how the patent system was administered and how and why the courts reached decisions. Administrative and judicial changes frequently have important economic implications, and these should not be ignored. Moreover, it is crucial to understand what contemporaries thought of the patent system and why they proposed changes. Their perception of the costs and benefits of patents will, after all, have determined what action they took to protect inventive output, and this in turn would have further implications for the rate of innovation and diffusion. Secondly, since technological change is a complicated social process, it frequently requires an interdisciplinary approach. In fact, Rosenberg has argued that research upon the subject must, necessarily, be so.[41]

Part two will assess the determinants of inventive activity and how far patents were important for inventors and the users of inventive output. Chapter 6 will analyse the patent data to show that inventive activity was primarily an economic activity. This will be supported in Chapter 7, where the trade in invention is examined. Chapter 8 will look at investment in invention, and the final chapter will examine the problems imposed by the changing quality of patent law.

The general intention, then, is to see what effect patents had on those who used the system. A number of important questions will be ignored. There will be no direct discussion of the net economic effects of the system as a whole, or what effect patents had on the development of industry generally. Agriculture is almost completely ignored, and nothing will be said about foreign or colonial patents. Nor will there be any attempt to assess whether inventions were labour or capital-saving, or what effects patents had on the rate of diffusion. These are all interesting questions, but would require separate studies. The purpose here is simply to analyse one aspect of the patent system and to provide an overall picture of the institutional framework within which the market for inventive output operated.

**Notes**

1 See A. E. Musson's excellent and widely ranging survey in *Science, Technology and Economic Growth in the Eighteenth Century*, 1972, especially pp. 49–56.
2 A. E. Musson and E. Robinson, *James Watt and the Steam Revolution*, 1969; D. McKie and E. Robinson, *Partners in Science: Letters of James Watt and Joseph Black*, 1969; E. Robinson, 'James Watt and the law of patents', *Technology and Culture*, 1971, pp. 115–39; H. W. Dickinson, 'Richard Roberts, his life and inventions', *Transactions of the Newcomen Society*, 1945–47, pp. 123–37; T. S. Ashton, 'Some statistics of the industrial revolution in Britain', *Manchester School*, 1948, pp. 214–34.
3 A. A. Gomme, *Patents of Invention: Origins and Growth of the Patent System in Britain*, 1946; H. Harding, *Patent Office Centenary*, 1953.
4 *Ibid.*, p. 147.
5 The number of patents for 1852 have been excluded because the system was reformed in October of that year. These data also exclude patents taken out for Scotland and Ireland, because there is no equivalent time series.
6 W. H. Price, *The English Patents of Monopoly*, 1906; J. W. Gordon, *Monopolies by Patents*, 1897; J. E. Neale, *Elizabeth I and her Parliament*, 1957; K. Boehm and A. Silberston, *The British Patent System: Administration*, Cambridge, 1967; C. T. Taylor and A. Silberston, *The Economic Impact of the Patent System: a Study of the British Experience*, Cambridge, 1973.

7 S. Smiles, *Industrial Biography*, Newton Abbot, 1967 ed., and *Lives of the Engineers: Early Engineering*, 1904; R. Davis, *Foreign Trade during the Industrial Revolution*, Leicester, 1979; R. M. Hartwell, *The Causes of the Industrial Revolution*, 1967.
8 P. Uselding, 'Studies of technology in economic history', in R. E. Gallman (ed.), *Recent Developments in the Study of Business and Economic History: Essays in Memory of Herman E. Krooss*, J.A.I. Press, 1977, pp. 159–219.
9 See G. F. Ray's review of Silberston and Taylor's book, *Economic Journal*, 1974, p. 443.
10 Robinson, *op. cit.*, 1971, p. 115.
11 Boehm and Silberston, *op. cit.*, p. 37.
12 T. S. Ashton, *The Industrial Revolution, 1760–1830*, Oxford, 1968, p. 11.
13 D. S. Landes, *The Unbound Prometheus*, Cambridge, 1970, p. 41. It is important to note, as will be shown later, that the cost of patenting did not alter during the industrial revolution.
14 W. Holdsworth, *History of English Law*, 1938, XI, p. 432.
15 H. G. Fox, *Monopolies and Patents*, Toronto, 1947, pp. 85, 212.
16 A. F. Ravenshear, *The Industrial and Commercial Influence of the English Patent System*, 1908, p. 55.
17 Harding, *op. cit.*, p. 4.
18 H. S. Hatfield, *Inventions and their Use in Science Today*, 1939, p. 175. Even though the origins of the patent system are obscure, it was certainly not a British invention. See W. Hamilton, 'Origin and early history of patents', *Journal of the Patent Office Society*, 1936, pp. 19–34; M. Frumkin, 'The origin of patents', *Journal of the Patent Office Society*, 1945, pp. 143–9.
19 J. Robinson, *The Accumulation of Capital*, 1969 ed., p. 87.
20 K. Arrow, 'Economic welfare and the allocation of resources for invention', in N. Rosenberg (ed.), *The Economics of Technological Change*, 1971, pp. 164–81; W. D. Nordhaus, *Invention, Growth and Welfare: a theoretical treatment of technological change*, Massachusetts, 1969; A. Silberston, 'The patent system', *Lloyds Bank Review*, 1967, pp. 32–44; H. G. Johnson, 'Patents and licenses as stimuli to innovation', *Weltwirtschaftliches Archiv*, 1976, pp. 420–8.
21 N. Rosenberg, *Technology and American Economic Growth*, New York, 1972, p. 188. Rosenberg includes as a restriction 'those inventions which would have been made even without a patent system, but which fall equally under its umbrella of protection'.
22 G. Prosi, 'Patents and externalities', *Zeitschrift für Nationalökonomie*, 1971, pp. 68–80.
23 A. Plant, 'The economic theory concerning patents for invention', *Economica*, 1934, pp. 30–51.
24 W. Kingston, *Invention and Monopoly*, W.E.P., 1968, p. 10.
25 F. M. Scherer, 'Nordhaus's theory of optimal patent life: a geometric reinterpretation', *American Economic Review*, 1972, pp. 422–7; W. D. Nordhaus, 'The optimal life of a patent: reply', *American Economic Review*, 1972, pp. 428–31.

## 12  Introduction

26 G. Prosi, 'Socially optimal patent protection — another exercise in utopian economics?', in J. S. Dreyer (ed.), *Breadth and Depth in Economics*, San Diego, 1979, pp. 271–9.
27 F. Machlup, *An Economic Review of the Patent System*, Study No. 15 of the Sub-committee on Patents, Trademarks and Copyrights of the Committee on the Judiciary, U.S. Senate 85th Congress, 2nd Session, Washington, 1958, pp. 79–80. Whilst economists are unable to provide a basis for choosing between 'all or nothing', Machlup suggests that with sufficient research they will be able to make decisions about 'a little more or a little less' patent protection.
28 E. Schiff, *Industrialisation without National Patents: the Netherlands, 1869–1912, Switzerland, 1850–1907*, Princeton, 1971.
29 One major criticism is that Schiff is inconsistent on what he means by the volume of inventive activity. In his sketch of inventive activity he examines '*great* Swiss inventions', when he earlier states that 'we are concerned mainly with the ability of a patent system to stimulate the propensity or eagerness to invent regardless of how important the inventions will prove'. The quality and quantity of inventions are two quite different things, as he himself admits. Industrial development can also, of course, be explained by other factors apart from invention and technological change. As is clear, differential historical analysis itself poses a number of conceptual difficulties.
30 See, for example, Samuel Crompton's letter to Joseph Banks. 'To secure to myself the benefits of my own labours that I am desirous to obtain, I have been told that a caveat might be put in *somewhere* that would secure to us our right so that in some future day we might take out a patent if it was necessary. But how to go about this, I do not know, nor who is the most qualified to direct me.' Egerton MSS, 2409, Oct. 1807, British Museum. See also H. C. Cameron, *Samuel Crompton, 1753–1827*, 1952, pp. 94–5. It seems unlikely that many were as ignorant as Crompton, but it took time for some to understand the advantages of patenting. See H. Bessemer, *Autobiography*, 1905, p. 25; *Select Committee on Patents*, Parl. Papers, XVIII, 1851, p. 286.
31 See, for example, S. Lilley's attempt to rank the quality of inventive output in *Machines and History*, 1948. His method of giving marks out of ten was sensibly dropped for the second edition.
32 For a discussion of patent statistics see J. Jewkes, D. Sawers, R. Stillerman, *The Sources of Invention*, 1969, 2nd ed., pp. 88–91, and below.
33 J. Schmookler, *Invention and Economic Growth*, Cambridge, Mass., 1966, p. 56.
34 For a review of this vast literature see F. M. Scherer, *Industrial Market Structure and Economic Performance*, Chicago, 1971, pp. 346–99; J. V. Koch, *Industrial Organisation and Prices*, 1980, pp. 213–50; E. A. Mansfield, *The Economics of Technological Change*, 1969, and *Industrial Research and Technological Innovation*, 1968.
35 Jewkes, *op. cit.*, p. 25.
36 W. R. Maclaurin, 'The sequence from invention to innovation and its relation to economic growth', *Quarterly Journal of Economics*, 1953, pp. 97–111.

37 Usher divides the act of invention into four stages: (1) perception of the problem, (2) preparation and setting, (3) the primary act of insight, (4) critical revision and development. See A. P. Usher, *A History of Mechanical Inventions*, Cambridge, Mass., 1954, p. 60, and 'Technological change and capital formation', in M. Abramovitz (ed.), *Capital Formation and Economic Growth*, Princeton, 1956, pp. 533–8; V. W. Ruttan, 'Usher and Schumpeter on invention and innovation and technological change', *Quarterly Journal of Economics*, 1959, p. 599; W. P. Strassman, *Risk and Technological Innovation*, Ithaca, 1959, p. 9.

38 See for example, J. E. S. Parker, *The Economics of Innovation: the national and multinational enterprise in technological change*, 1974, p. 36.

39 For a different view see C. S. Solo, who argues that during the industrial revolution both processes were performed by the same person. Only with the growth of patents were these two activities performed by separate people. 'Innovation in the capitalist process: a critique of the Schumpeterian theory', *Quarterly Journal of Economics*, 1951, pp. 417–28.

40 J. Schmookler, *op. cit.*, 1966, ch. 1. See also S. Kuznets, 'Inventive activity: problems of definition and measurement', in R. R. Nelson (ed.), *The Rate and Direction of Inventive Activity*, Princeton, 1962, pp. 20–4.

41 N. Rosenberg, *Perspectives in Technology*, Cambridge, 1976, p. 83. He continues thus: 'The exhortation to undertake interdisciplinary research, we all know, is a familiar one — just as we also know how infrequently it has been undertaken successfully. The crossing of disciplinary boundaries is likely to be a hazardous operation ... [and] ... It will be accomplished, not in response to mere exhortation, but only by the eventual realisation that certain problems of major significance to practitioners of one discipline will resist satisfactory resolution unless there is a willingness — and a capacity — for crossing such lines whenever the intellectual chase demands it.'

# PART I

The patent institution

# 1

# The case for the patent system

During the industrial revolution the patent system was justified by four arguments: the natural-law thesis, the reward-by-monopoly thesis, the monopoly-profit thesis, and the exchange-for-secrets thesis.[1] The 'natural rights' thesis was the least important and was practically abandoned by the late 1820s, but the remaining three were widely advanced before and during the patent controversy of the mid-Victorian period. This chapter will consider these arguments and assess the rise of the anti-patent movement. It will be shown that the arguments for the abolition of the patent system are rarely encountered until *after* the 1852 Patent Reform Act, despite the often vigorous opposition to a number of individual patents. During the industrial revolution the theories supporting the patent *system* went largely unchallenged. Most endorsed the economic philosophy expressed in the 1624 Statute of Monopolies, which, as one writer noted, provided the 'first germ of the Patent Law, springing forth from the destruction of despotic privilege, like the young tree from the ruined feudal castle'.[2]

The four arguments justifying patent protection were in general circulation in the early nineteenth century. They were put forward by a wide variety of individuals, but principally by political economists, lawyers, engineers, patent agents, inventors and manufacturers. Large tomes, pamphlets, petitions, private correspondence and public statements made before the two Select Committees on Patents testify to the universality of these views.

The natural-law thesis assumes that individuals have a natural right of property in their own ideas. On this argument, using the ideas of others without some form of compensation amounted to theft, and since property was personal and exclusive, the State was morally obliged to enforce exclusivity. This reasoning was enshrined in the French patent law of 1791 and openly advocated by J. R. McCulloch during the patent reform campaign of the 1820s: 'If anything can be called a man's exclusive property, it is surely that which owes its birth entirely to combinations formed in his own mind, and which, but for his ingenuity, would not have existed.'[3] Despite believing that government intervention with the methods of manufacturing was likely to check invention, McCulloch continued to urge that patents would increase the level of inventive activity without harming the public. As a general rule, though, the natural-law theory of property in inventions was rarely advanced by supporters of patents.[4] When the *Westminster Review* brusquely announced that 'to talk of the natural rights of an inventor is to talk nonsense',[5] it was condemning an idea already widely discredited. Occasionally some writers would resort to the argument, but no worthwhile commentator took it seriously.[6] By the time J. E. T. Rogers was criticising the thesis in 1863 it had been abandoned.[7] In law at least, the matter was clear. William Hindmarch, the leading academic patent lawyer, put the case succinctly.[8] 'No inventor can, in fact, have any natural right to prevent any other person from making and using the same or similar invention, and therefore the law does not recognise any right or property whatsoever in an invention which is not made subject to a grant by patent.'[9] Thomas Webster, an engineer, leading patent lawyer and patent reformer, was equally definite when he argued that patent monopolies were only temporary, because inventors had no natural or inherent right in their inventions. 'Those who believe the inventor to have a natural right ... must have an entire misconception as what it is the inventor really achieves.'[10] On the whole, it seems that the natural-rights-in-invention thesis was more extensively employed by Continental countries, and consequently it is there that most of the critical writing appears.[11]

The reward-by-monopoly thesis, in contrast, was widely used to justify patents. It was based on the notion that inventors should be rewarded according to the usefulness of their invention. Since this reward cannot be guaranteed by ordinary market forces, the State should intervene to provide a temporary monopoly. Adam Smith, more than most other political economists, was careful to appreciate that the happy miracle of economic harmony was not simply created out of chaos by the invisible hand.[12] The law and lawmaker were crucially important in ensuring competition and the most efficient allocation of resources. For Smith the institutional framework, which included legal and non-legal elements, was logically prior to the market. He therefore supported patents for inventors, and for two related reasons. Firstly, patents were justified because they allowed inventors a monopoly period during which they could appropriate a sufficient return for their ingenuity and effort. Without this kind of protection, competitors would be able to reduce their own costs without bearing any of those involved in producing inventive output, and it would not pay individuals to invent because of those who gained a free ride, that is, benefited without paying the cost.[13] Secondly, patents were justified because they were harmless. Like every other commodity, inventive output was subject to the laws of supply and demand. 'For here, if the invention be good and such as is profitable to mankind, he [the inventor] will probably make a fortune by it; but if it be of no value he will reap no benefit.' As far as Smith was concerned, 'if the legislature should appoint pecuniary rewards for the inventor ... they would hardly ever be so precisely proportioned to the merit of the invention as this is'.[14] Despite his views on exclusive privileges and monopolies, Smith was clearly in favour of the economic benefits conferred on inventors by patents.

Bentham also justified patents principally because they provided the inventor with a reward for his efforts.[15] He came to this conclusion by distinguishing between two categories of labour. Firstly, there is the physical kind of labour which merely produces goods. When imitated, this kind of labour requires an equal amount of labour to produce an equivalent output: the inputs per unit of output are the same. The second kind of labour is qualitatively different and is defined by the skill or mental power used in the production of goods. The imitation of this kind of labour leads to relatively lower input costs because, while knowledge is expensive to produce, it is inexpensive to *re*produce: 'of skill ... it is the property to be capable of being indefinitely imbibed and diffused and that without any exertion of mental labour comparable to that, at the expense of which it was acquired'.[16] This is the free-rider problem elliptically expressed. 'A man will not be at the expense and trouble of bringing to maturity [an] invention unless he has a prospect of an adequate satisfaction, that is to say, at least of such a satisfaction as to his eyes appears an adequate one, for such troubles and expense.'[17] Bentham never doubted that patents, compared with any other system of encouraging and protecting invention, were 'proportionately and essentially just'.[18]

John Stuart Mill was equally certain that an exclusive privilege of a temporary duration was the most efficient means of rewarding inventors and, like Smith, he wanted the market to determine what the reward should be; 'the reward conferred by ... [patents] ... depends upon the invention's being found useful, and the greater the usefulness the greater the reward'. Though Mill admitted that the patent laws as they stood in 1848 were in need of some improvement, the principles upon which patents were granted were incontestable: 'it would be a gross immorality in the law to set everybody free to use a person's work without his consent and without giving him an equivalent'.[19]

For the classical economists patents were not a burning issue; their observations were usually expressed only as part of larger arguments concerning monopolies, the role of government and the consequences of the division of labour.[20] Contemporary inventors were, naturally, far more enthusiastic. In 1774 W. Kendrick argued that the 'most plausible and politic method of bestowing that encouragement is therefore, that by which the eventual utility of such invention is made the measure of reward. This is effected by letter patent.'[21] A similar homily was published by an anonymous observer in 1791: 'Patents are the greatest and the most effectual incitement that can be contrived ... to recompense ingenuity.'[22] In Hornblower v. Boulton one counsel argued, 'every new invention is of importance to the wealth and convenience of the public, and when they are enforcing the fruits of a useful discovery it would be hard upon the inventor to deprive him of his reward'.[23] Many MPs shared this view. T. Lennard considered that the patent system was sound in principle because the inventor's reward depended on the value of the invention itself.[24] Luke Herbert repeated these views in his *Engineers' Encyclopaedia*, published in 1838,[25] and by the 1840s and 1850s this view provided a stock argument for every writer on the subject.

The monopoly-profit-incentive thesis was probably the most quoted argument in support of patents. Private reward, it is clear, can also act as an incentive to invent, but the reward argument is concerned with some just profit which the inventor in some sense deserves to make within the monopoly period. The incentive argument is concerned with the duration and exclusiveness of the monopoly itself and is linked with the notion that economic growth is inherently desirable. It assumes that the supply of invention (which was also assumed to be a major cause of growth) would be less than it would otherwise be if patents were not used to protect the inventor. Here inventive activity was associated with progress as well as private profit, and this probably explains why the argument was so popular during the early nineteenth century.

There is ample evidence of the belief in the causal link between

# The case for the patent system

patents, invention and industrial development. In 1791 Sir William Pultney could write to Lord Kenyon that 'I think [patents] have been one of the great causes of the important discoveries which in this country have so much improved our manufactures and trade'.[26] Another writer held that patents encouraged 'some of the most valuable inventions which the various and astonishing powers of mechanics have produced. If new inventions,' he concluded, 'are not protected, England's sun is set ... [and] ... the mechanical genius of this country will sleep'.[27] John Chitty wrote in a similar vein that patents to the 'first inventor' were the most effective means of encouraging the 'production of GENIUS'.[28] Witnesses examined by the 1829 Select Committee on Patents reiterated these views. W. H. Wyatt, editor of the *Repertory of Arts*, believed that the patent system was the greatest 'spur to the improvements of the arts and manufactures in this country'. The engineer John Farey, who had a vast experience of invention and inventors, argued that 'in all cases, an invention is more speedily brought to perfection under a patent than without, and in most cases it is more speedily brought into general use'.[29] Patents, he told Lord Brougham, were vitally important for natural prosperity.[30] By the 1830s it was scarcely necessary to suggest 'that facilitating the acquisition of a patent was amongst the most effective modes of advancing the best interests of society'.[31]

Foreigners were equally impressed. J. C. Fischer noted in his diary that 'the English did not fear foreign competition in [mechanical arts] as in other branches of industry. The system of granting patents ... protected and stimulated the industrial economy and probably helped secure the production of manufactured goods'. As far as the editors of the burgeoning number of journals and magazines associated with invention and patents were concerned, the links between patents, inventive activity and prosperity were obvious.[32] When the patent reform movement renewed agitation in the years before the Great Exhibition, innumerable petitions requesting greater protection for inventors extolled — and for equally obvious reasons — the virtues of patents. The petition from the Association of Patentees for the Protection and Regulation of Patent Property claimed typically that the 'unequalled progress which the useful arts have made in Great Britain, and the national opulence and greatness of which they have been confessedly the prime source, are distinctly traceable to the encouragement afforded to inventors by the patent law'.[33] Witnesses examined by the 1851 Select Committee on Patents continued to emphasise these links between patents and prosperity. Many agreed with William Carpmael that the 'manufactures of this country would not have been anything like what they are, had it not been for the patent laws'.[34] Legislative interference was absolutely necessary for the protection and encouragement of invention, industry and trade.

The final justification for patents was the exchange-for-secrets thesis, or the disclosure agreement. It was based on the eighteenth-century idea of contract, where society and the inventor made a bargain, one offering temporary protection, the other knowledge of new techniques. Unlike the incentive argument, disclosure was not related to the supply of inventive output. It was merely concerned with the dissemination of information of *existing* technology which would otherwise remain secret. It did, however, assume — as did the incentive argument — that invention and progress were causally linked.

This rationale had its origins in the Elizabethan period, although then disclosure was of a quite different form. Inventors were compelled to use the patent to introduce the trade, and to teach the mystery of the art to native tradesmen. In the early eighteenth century the form and condition of disclosure changed. Patentees now had to describe the nature and manner of their inventions in a specification. In 1734 the law officers made it a condition of the contract by a provision in the patent itself, and in the 1778 Liardet and Johnson case Lord Mansfield gave this form of contract a secure legal basis. This was confirmed by Buller, J., in King *v.* Arkwright (1785) and in Williams *v.* Williams, Lord Eldon held that patentees were 'purchasers from the public': patents were exchanged for secrets. Lennard told Parliament in 1829 that patents 'must not be confused with the common notion of monopoly, which it was not, being merely a bargain between the inventor and public'.[35] John Farey put the matter simply; a 'patent is the price of disclosure'.[36] For the *Mechanics Magazine* the 'only ground on which it can be considered good for the community at large to encourage the taking out of patents, is that it may cause many new and useful inventions to be made public which might otherwise be lost for ever'.[37] Many, in fact, considered that disclosure was analytically the mose important ground for supporting patents, and by 1851 the argument was commonplace.

When putting these arguments forward great care was taken to explain in what sense patents were monopolies. Coke's dictum 'monopolies in times past were ever without law but never without friends'[38] had a haunting ring. As one letter writer to the *Mechanics Magazine* observed, 'it is a misfortune, and not a small one, that patents for inventions have descended historically as arbitrary grants of privilege ...

and this erroneous view ... has not been entirely discarded to this day'.[39] To forestall opposition the supporters of patents used three arguments. Firstly, patents did not deprive the public of anything which they had previously enjoyed. The inventor, by definition (if not always in practice) made and produced either a new process or a new product. 'The ultimate test,' wrote Webster,

> by which the character [of a monopoly] may be ascertained is, whether the monopolist benefits himself *without* injuring others ... with a monopoly of *existing* trade all others are excluded from that which the public were already in the possession and practice of, which is a clear violation of public right and policy, but with the manufacture of a new invention ... nothing is taken from the public which it before possessed.[40]

As early as 1775 Edmund Burke had argued much the same, though in less legalistic language: 'Monopoly is an odious term ... [but a patent] ... is not making a monopoly of what was common. It is the direct reverse, for the condition of the patent, compelling a discovery, makes that common which was private before.'[41]

Secondly, patents could not be considered as monopolies because they did not lead to artificial increases in prices. Pre-expiration prices were simply higher than their post-expiration prices. It was not accurate, therefore, to claim that patents increased existing prices: the article produced must be better (and therefore qualitatively different) or cheaper. 'A patent is recompense to the patentee at the expense of the consumer; *but* the tax is a willing tax with a short duration.'[42] Lastly, a few suggested that all monopolists were in competition with all other producers. This quite modern conception of competition was argued in particular by William Spence. 'I think that a patent is a peculiar kind of monopoly because it is not a monopoly in every sense of the word, for a patentee is in competition with all others in the open markets of the world.'[43] As early as 1803 these arguments had persuaded one writer to state that 'this poison [patent monopolies] has been deprived of all its pernicious ingredients and converted into a nutritious aliment applicable to the support of commercial prosperity'.[44] Inventors were no longer ashamed of being patentees and, in fact, were proud of being known as such.[45]

These arguments were not unopposed, and by 1869 *The Economist* proudly predicted that 'it was probable enough that the Patent Laws will be abolished ere long'.[46] *The Times* and the *Spectator* likewise anticipated that patents would soon be 'wiped out of the statute book'.[47] These confident predictions were made when the patent abolition movement was at its peak, and when the fervour with which the movement expressed its hostility to patents resembled the anti-patent debate of the early seventeenth century. Now the direction and content of the argument were largely different. The Crown's abuse of the royal prerogative and its use of patents as a source of patronage and revenue were no longer the issues of contention. In the 1860s opposition turned on the restraining effects that patents had on industry and enterprise. But not all links with the past were broken. John Chamberlain's protest in 1620 'that patents are become so ordinary, that there is no end, every day bringing forth some new project or other'[48] would have been sympathetically endorsed by all mid-Victorian abolitionists. For the supporters of patents the past provided another lesson. The 'destroyers of patent protection', wrote Henry Dircks, 'form a class which may well rank with the bygone destroyers of machinery'.[49]

The main ideological influence behind the anti-patent movement was free trade and the emancipation of industry. 'Like almost every ... defence of patents,' Robert Macfie wrote, 'it ignores the grand objection ... their incompatibility with Free Trade.'[50] This theme was repeated in every pamphlet and speech urging abolition, and after the repeal of the corn laws and Navigation Acts the argument had some force. References to the invidious effects of monopoly, the hardships imposed on domestic manufactures, and the debilitating effects this would have on the export trade, all served to evoke memories of the anti-corn law campaign. Patents merely clogged the wheels of progress, and prevented manufacturers eager to improve efficiency from using inventions unless they were prepared to pay for unnecessarily expensive licences.

Another argument which grew out of this free-trade mood was the notion that patents were no longer necessary for an industrialised and mature economy. Firms and industries were now able to compete efficiently without special privileges. Britain was the workshop of the world, and no longer needed to rely upon the old instruments of protection. Patents were simply redundant. 'My idea,' Cubitt informed the 1851 committee,

> is that in a rude and infant state of society where there is a very great room for improvement, patents may be, if not good, better than they are in an improved state of society; the more improved society becomes and the more perfect is the application of science particularly to the ends in view, the less can patents be supported, and the less value they really are.[51]

F. Edwards shared this early Victorian optimism. 'The patent system has in its time answered a good purpose ... but now matters are altogether altered. Enterprise is so great that infinitely greater things are doing without special privileges than with.'[52]

For some abolitionists patents were not only unimportant but positively harmful, both morally and economically. In 1851 *The Economist*, then in rampant mood, urged that the granting of patents 'inflames cupidity, excites fraud, stimulates men to run after schemes ... begets disputes and quarrels betwixt inventors, provokes endless lawsuits [and] makes men ruin themselves for the sake of getting the privilege of a patent, which merely fosters a delusion of greediness'.[53] Brunel, in contrast, dressed up his argument in the language of the economist. Patents were harmful because they diverted inventive activity into channels which wasted the resources of inventors better employed in their ordinary occupations. The injurious effects that patents had on inducing silly speculations and nonsensical schemes was a point also taken up by Lord Granville: 'patents gave a factitious stimulus to false invention'.[54] In short, the social costs of patents exceeded any benefits which the system created.

Most abolitionists did not seriously challenge the argument that inventors should receive some reward, although there were some exceptions. William Armstrong suggested that they needed no financial spur, since they could not help but invent. Invention was the product of the instinct of contrivance, not the demanding needs of maximising profits.[55] *The Economist* argued that the inventors' head-start over competitors would generate sufficient reward on its own.[56] J. L. Ricardo went much further. In his view, invention was the result of impersonal forces rather than the consequence of varying degrees of individual genius. Since nearly all useful inventions depend less on any individual than on the progress of society, 'there was no reason to reward him who might be lucky enough to be the first to hit on the thing required'.[57] A number — even supporters of patents — agreed that the patent system was little more than a lottery and frequently 'bestowed rewards *on the wrong* persons',[58] but few went so far as to deny inventors financial return for their labours. Most abolitionists were principally concerned with the *form* of reward.

The idea that inventors should be rewarded by a different method was an old one. In the eighteenth and early nineteenth centuries the Society of Arts sought to increase the supply and nature of inventions through the use of premiums and medals, but with little success.[59] The few awards made by Parliament were also generally inadequate, whilst the myriad proposals for Boards of Invention never got off the ground because there was no way of assessing the monetary value of an invention *ex ante*.[60] Patents at least let the market decide. 'Honours, rewards and medals', as Charles Babbage scathingly noted, were nothing more than the 'feeble expression of the sentiments of mankind'.[61]

These arguments have encouraged some writers to believe that the principles of the patent system were challenged at the very beginning of the industrial revolution. W. Bowden points to the intense opposition to patents in the late eighteenth century, especially in the textile areas of northern England.[62] T. Daff suggests that the move towards reconstructing the system during the early nineteenth century 'sparked off a heated debate concerning whether such a system was really necessary'.[63] Boehm and Silberston hold a similar view and quote the opposition to Arkwright's patent and to Watt's 1775 patent extension as evidence for their argument.[64] The 1820 petition sent to Parliament by several patentees, engineers and manufacturers, together with one made in 1829, in which it was claimed that the evils of patent systems were 'inconsistent with the maxims of sound political economy',[65] further suggests that the case for abolishing patents was continuously urged throughout the period 1780 to 1850. There is, though, little evidence to support this view. Except for the attitudes of some judges (which will be discussed in another chapter), abolitionist views are rarely encountered during the classic period of the industrial revolution. What at first appears to be a serious challenge to the principles of the patent system frequently turns out to be something quite different, usually a demand for reform.

The most common cause of protest was the *inefficient* way in which the patent system was administered by the courts and the Patent Office itself. Many inventors, bitterly disappointed with the protection which patents offered, frequently condemned them as useless and often threatened not to patent further inventions until the system was reformed. Josiah Wedgwood's letter to Lord Dundonald is a quite typical late eighteenth-century view.

> I am not surprised at your Lordship's aversion to patents. *They are bad, and deficient for the purpose intended in many respects*, and as any foreigner may learn the discoveries for which patents have been granted at the expense of a few shillings and practice them immediately in other countries whilst the hands of all British artists and manufacturers are bound during the term of the patent. Considered in this light patents are

highly pernicious to the community amongst whom the invention originated and a remedy is much wanted in the Patent Office for this evil.[66]

Here it is clear that Wedgwood is simply concerned to show that patents offered inadequate protection against foreign competition because patent specifications were disclosed to the public. In the nineteenth century, as the fear of overseas competition diminished — except in depressed years — inventors switched their criticisms to the inadequate protection offered against domestic manufacturers. Dr Cartwright provides another typical example. 'A patent is a feeble protection against the rapacity, piracy, and theft of too many of the manufacturing class. There is scarcely an instance, I believe, of a patent being granted for any invention of a real value, against which attempts have not been made to overthrow or evade it.'[67] The failure of patents to protect inventions adequately became a major concern with inventors and manufacturers in the first half of the nineteenth century, and led to a vociferous reform, not abolitionist, movement.

Another important reason for protest was the 'rapacity' of inventors themselves, especially those who intended to do more than they were legally entitled to do. In this case opposition was against particular patents, not against the patent system itself — two quite separate issues. The massive opposition to Arkwright's patent was based on his 'tyranny' of the trade, not on any strong desire on the part of the Manchester men to abolish patents. The Manchester Commercial Committee, which had been reorganised on a wider basis in 1781 and 1782, reported a wish to devise alternative forms of rewarding inventors, but the idea seems to have died after Arkwright's patent was set aside in 1785.[68] Nor is it altogether certain that they wished to abolish the patent system.[69] When the *Manchester Mercury* gleefully announced the cancellation of Arkwright's patent as 'liberating the country from the dreadful effects of a monopoly in spinning'[70] the argument was not extended into general criticism of the system as a whole. This is also the message in Gravenor Henson's attack on John Heathcote's patent, even though the language may suggest otherwise. Patent law is 'now ... an evil sneaking in a corner which works in secret and is used only as an instrument of private oppression ... [and] it has been of late years very seldom devoted to the encouragement of the man of genius or the real inventor; but has been the engine wherewith the rich man has frequently and too generally oppressed a whole district and has acquired a princely fortune by purse and terror alone'.[71] In many ways critics like Henson (a supporter of patents) were misleading and often gave the impression of attacking the whole system when in fact they were really doing nothing of the kind.

Finally, protest seems to be directed less at patents than at patent extensions. The petition from Liverpool opposing Arkwright's patent is quite clear on this issue. Arkwright has

> obtained every advantage which the laws in being have provided for the Encouragement of new Inventions, and, by the Emoluments arising from this exclusive grant, hath, as the Petitioners are informed, and believe, realized such a fortune as every unprejudiced Person must allow to *be an ample Compensation for the most happy efforts of Genius*; and that the Interposition of the Legislature to grant still great Rewards, and such as are unwarrantable by any statute now in being, might as the Petitioners apprehend give stability to a dangerous Monopoly, and prove highly prejudicial to the manufacturers, commerce and navigation of this land by leaving to all other Kingdoms the unlimited Use of Machines which would in England [be denied].[72]

Despite all the special pleading, there is no critique of the existing law as embodied in the Statute of Monopolies: patents were acceptable, whereas patents renewals were not. Manchester's opposition to Anthony Bourboulon de Boneuil's and Matthew Vallet's petition to prolong their bleaching patent is further evidence that extensions rather than patents *per se* were the source of concern.[73] And when Samuel Morton petitioned to extend his patent in 1832, the chairman of the Select Committee investigating the claim observed that there was a strong feeling against applications for the renewal of patents because it altered the contract between inventors and the public. In saying this, though, he did not imply that patents themselves were an evil which ought to be abolished.[74]

The public case for abolition was in fact almost wholly confined to the late 1850s and 1860s. This is confirmed by the leaders of the movement themselves. 'It is in fact only within a recent period that patents for invention have been strongly attacked,' wrote F. Edwards in 1865.[75] Lord Granville, who was chairman of the 1851 Select Committee and an ardent abolitionist — a matter which did not comfort patent reformers — told Parliament he would have no difficulty in finding a hundred sensible people to support the necessity of a patent law, 'but with respect to the injurious tendency of the whole system there were probably not six persons who could be got to give evidence in support of that view'.[76] R. A. Macfie, sugar refiner and the leading figure in the movement, reluctantly agreed, and even as late as 1864 he was not confident that abolition would be achieved.[77] The

## The case for the patent system

*Mechanics Magazine*'s fear that the 1851 committee was conspiring to abolish patents was, consequently, unjustified.[78] Only Granville, William Cubitt, John Fairre, Robert Macfie, J. L. Ricardo (nephew of David Ricardo), Romilly (the Master of the Rolls), J. H. Lloyd and Brunel preached abolition. Fairbairn, Maudsley, Field and Penn were alleged to be averse to patents, but not owning them.[79] Outside the 1851 committee, support was no greater. John Scott Russell, who pleaded guilty to having three patents, now wanted to see property in patents 'annihilated', as did E. B. Denison from Leeds.[80] In 1851 the sugar refiners of Liverpool, Greenock, Glasgow and Leith also petitioned to abandon the system, but they were no doubt strongly influenced by the two leading refiners, Macfie and Fairre. In fact sugar refinding was the only industry which advocated abolition. Otherwise only *The Economist*, with its uniquely distilled versions of *laissez-faire*, gave the infant movement any support at a national level.[81] By the late 1860s and early 1870s, though, the anti-patent movement had become a force of some strength. The fact that Switzerland and Holland had abolished their patent systems in 1863 and 1869 gave the British movement an impetus which it never previously had. During the industrial revolution scarcely anyone had wanted the patent system abolished.

With the emergence of protectionism during the Great Depression the anti-patent movement collapsed as suddenly as it appeared. Intense international rivalry now made the patent system perfectly respectable once more. Expediency, though, was not the only explanation. The supporters of patents had history on their side. They could point to the rapid progress of the last hundred years, and to the equally rapid improvements in technology, protected by patents. The unenviable task of the abolitionists was to show that there was no causal relationship between patents and progress, and this they understandably failed to do. The very existence of the system was its greatest advantage, and despite its faults there were no suitable alternatives. Matthew Davenport-Hill summed the matter up philosophically:

> That the wit of man cannot devise a perfect [patent] system — perhaps not one approaching perfection — I do not deny. But every day I live, the more strongly am I impressed with the belief that this is and ever was and ever will be the condition of all human affairs; and that we must be satisfied with very distant approaches indeed to what it is very desirable we should attain.[82]

## Notes

1. F. Machlup and E. Penrose, 'The patent controversy in the nineteenth century', *Journal of Economic History*, 1950, pp. 1–29; Machlup, *An Economic Review of the Patent System*, op. cit., pp. 1–87.
2. T. Turner, *Remarks on the Rights of Property in Mechanical Invention with remarks to Registered Designs*, 1847, p. 8. For earlier economic writers see C. Macleod, *Patents for Invention and Technical Change in England, 1660–1753*, unpublished Ph.D., University of Cambridge, 1982, pp. 315–63.
3. *Scotsman*, 26.5.1826; J. R. McCulloch, *Commercial Dictionary*, 1832, pp. 817–18; *The Literature of Political Economy*, 1845, p. 313; D. P. O'Brien, *J. R. McCulloch: a Study of Classical Economics*, 1970, p. 15.
4. H. D. MacCleod appears to have been the only other economist who accepted the natural-law view: *The Elements of Political Economy*, 1858, pp. 181–82. More generally see O. H. Taylor 'Economics and idea of *jus naturale*', *Quarterly Journal of Economics*, 1929–30, pp. 205–41; D. P. O'Brien, *The Classical Economists*, 1975, chs. 1–3.
5. *Westminster Review*, 'The patent laws', XXVI, 1829, p. 329.
6. R. Godson, *A Practical Treatise on the Law of Patents*, 1840, 2nd ed., pp. 1, 10; *Select Committee on Patents*, Parl. Papers, XVIII, 1851, p. 401.
7. J. E. T. Rogers, 'On the rationale and working of the patent laws', *Journal of the Statistical Society of London*, XXVI, 1863, p. 128.
8. *Mechanics Magazine*, XLIV, 1846, p. 300.
9. W. Hindmarch, *Law and the Practice of Letters Patent for Invention*, 1848, p. 228.
10. T. Webster, *Law and Practice of Letters Patent for Invention*, 1841, p. 3.
11. Machlup, op. cit., p. 22.
12. N. Rosenberg, 'Some institutional aspects of the *Wealth of Nations*', *Journal of Political Economy*, 1960, pp. 557–70; J. M. Buchanan, 'Public goods and natural liberty', in *The Market and the State: Essays in Honour of Adam Smith*, 1976, T. Wilson and A. S. Skinner, eds., pp. 271–95; L. Robbins, *The Theory of Economic Policy in English Classical Political Economy*, 1953; R. L. Crouch, '*Laissez-faire* in nineteenth century Britain: myth or reality?', *Manchester School*, 1967, pp. 199–215.
13. M. Olson, *The Logic of Collective Action*, Cambridge, Mass., 1965.
14. A. Smith, *An Inquiry into the Nature and Causes of the Wealth of Nations*, 1976 ed., R. H. Campbell, A. S. Skinner and W. B. Todd, eds., p. 593; A. Smith, *Lectures on Jurisprudence*, 1978 ed., R. L. Meek, D. D. Raphael and P. G. Stein, eds., pp. 83, 472. The treatment of patents is more explicit in the *Lectures* than in the *Wealth of Nations*.
15. W. Stark, ed., *J. Bentham's Economic Writings*, 1952 ed., I, p. 265.
16. *Ibid.*, p. 261.
17. *Ibid.*, p. 262.
18. *Ibid.*, p. 264.
19. J. S. Mill, *Principles of Political Economy*, 1902 ed., p. 563.
20. The Political Economy Club, which was formed in 1821 and which frequently discussed major contemporary issues, did not debate patents

until late 1854. J. L. Mallet ed., *Political Economy Club, Minutes of Proceedings, etc., 1821–1921*, 1921, pp. 69, 77.
21 W. Kendrick, *An Address to the Artists and Manufacturers of Great Britain Respecting an Application for the Encouragement of New Discoveries in the Useful Arts*, 1774, p. 20.
22 *Observations on the Utility of Patents ...*, 1791.
23 A. Prince, *The Law and Practice of Patents and Registration of Designs*, 1845, p. 20.
24 *Hansard*, XXI, 1829, pp. 598–608.
25 L. Herbert, *Engineers' Encyclopaedia*, 1838–39, p. 265.
26 Kenyon MSS, Sir W. Pulteney to Lord Kenyon, 12.5.1791. B.M. No. 1361.
27 *Observations, op. cit.*, pp. 14, 25.
28 J. Chitty, *A Treatise on the Law of Commerce, Manufactures and Contracts*, I, 1820–24, p. 6; *Times*, 22.4.1826.
29 *Select Committee on Patents*, Parl. Papers, III, 1829, pp. 103, 141.
30 Brougham MSS. [45, 493] Farey to Brougham, 21.8.1833, University College Library (hereafter U.C.L.).
31 *Hansard*, XXXVI, 1837, pp. 554–8.
32 W. O. Henderson, ed., *J. C. Fischer and his Diary of Industrial England, 1814–1851*, 1966, p. 41.
33 See *Mechanics Magazine*, LIV, 1851, pp. 9–12.
34 *Select Committee on Patents*, Parl. Papers, XVIII, 1851, p. 203.
35 *Hansard*, XXI, 1829, p. 601.
36 *Select Committee on Patents*, Parl. Papers, III, 1829, p. 21.
37 *Mechanics Magazine*, XIX, 1833, p. 297.
38 Cf. D. O. Wagner, 'Coke and the rise of economic liberalism', *Economic History Review*, 1935–36, p. 42.
39 *Mechanics Magazine*, XXV, 1836, pp. 399–401.
40 Webster, *op. cit.*, 1841, pp. 2–3; Hindmarch, *op. cit.*, p. 3; W. H. Wyatt, *A Compendium of the Law of Patents for Invention*, 1826, p. 2.
41 Burke MSS. 5923. An undated fragment. National Library of Ireland. Most of this note is illegible but it seems to have been written in 1775, when Burke was involved in petitioning Parliament to extend Richard Champion's patent. See also *The Correspondence of Edmund Burke*, III, 1961, G. H. Guttridge, ed., pp. 138, 142, 152–3, 156–9, 162–3.
42 T. Webster, *On Property in Designs and Inventions in Arts and Manufactures*, 1853, pp. 20–2; *Select Committee on School of Design*, Parl. Papers, XVIII, 1849, p. 392.
43 *Select Committee on Patents*, Parl. Papers, XVIII, 1851, p. 38; *Select Committee on Patents*, Parl. Papers, XXIX, 1864, pp. 385–6; *Times*, 6.6.1827.
44 J. D. Collier, *Law of Patents*, 1803, p. 12.
45 J. Davies, *Patent Cases*, 1816, p. 7.
46 *The Economist*, 5.6.1869.
47 *Times*, 29.5.1869; *Spectator*, 5.6.1869.
48 Price, *op. cit.*, p. 30.
49 H. Dircks, *Inventors and Invention*, 1867, p. 84.
50 R. A. Macfie, *Recent Discussions on the Abolition of Patents for Invention*, 1869, p. 3.

## The patent institution

51 *Select Committee on Patents*, Parl. Papers, XVIII, 1851, p. 456.
52 F. Edwards, *On Letters Patent for Invention*, 1865, pp. 66–7.
53 *The Economist*, 26.7.1851.
54 *Hansard*, CXVIII, 1851, p. 13.
55 *British Association for the Advancement of Science*, 1864, p. lii.
56 *The Economist*, 1.1.1851.
57 *Select Committee on Patents*, Parl. Papers, XVIII, 1851, p. 812.
58 *The Economist*, 1.1.1851.
59 D. G. C. Allan, *William Shipley, Founder of the Royal Society of Arts*, 1968, p. 43; T. H. Wood, *History of the Royal Society of Arts*, 1913, p. 243; *Select Committee on Arts and Manufactures*, Parl. Papers, V, 1835, p. 502; *Transactions of the Society of Arts*, LV, 1843–44, p. xv.
60 *Select Committee on Patents*, Parl. Papers, XVIII, 1851, p. 153; *Patent Journal*, I, 1846, p. 220; Brougham MSS, [3,974] J. Long to Brougham, 21.7.1835; [7,742], E. Hepple to Brougham, 31.10.1860, U.C.L.; P. Colquhoun, *Treatise on the Wealth, Power, and Resources of the British Empire*, 1815 2nd ed., p. 232.
61 C. Babbage, *Reflections on the Decline of Science in England and on some of its Causes*, 1830, pp. 132–3.
62 W. Bowden, *Industrial Society in England towards the End of the Eighteenth Century*, 1925, pp. 15, 29.
63 T. Daff, 'Patents as history', *Local Historian*, 1970–71, p. 277.
64 K. Boehm and S. Silberston, *The British Patent System*, 1967, pp. 24, 26.
65 *J.H.C.*, LXXV, 1820, p. 310; *J.H.C.*, LXXXIV, p. 339. This is not to suggest that some individuals did not *privately* oppose the patent system in principle. The 1820 and 1829 petitions for abolishing the system seem, however, to be the only two which were brought before the House of Commons in the period up to the early 1850s.
66 J. Wedgwood to Lord Dundonald, 26.3.1791, L — 17725 — 96, Wedgwood Museum.
67 Cartwright to Dr Bardsley, Sept. 1822; cf. E. Cartwright, *A Memoir of the Life, Writings and Mechanical Inventions of Edmund Cartwright*, 1971 ed., p. 295.
68 A. Redford, *Manchester Merchants and Foreign Trade, 1794–1858*, 1934, p. 5.
69 Redford seems to rely for his argument on the evidence provided by Bowden, who claims that the *'it appears'* that opposition to the *patent system* was one of the purposes of the 1782 committee, *op. cit.*, p. 167.
70 *Manchester Mercury*, 28.6.1785.
71 Add. MSS. 27,807. Henson to F. Place, 31.5.1825. For a different species, but equally misleading argument, see *Appendix 18th Report on Public Petitions*, 18,254, App. 15,114, 1848, pp. 812–13.
72 *J.H.C.*, XXXIX, 1783, p. 264.
73 Musson and Robinson, *op. cit.*, 1969, p. 285.
74 *Select Committee on Morton's Slip*, Parl. Papers, V, 1832, p. 297.
75 F. Edwards, *op. cit.*, p. 2; T. H. Farrer, *The State in Relation to Trade*, 1883, p. 60; Webster, *op. cit.*, 1853, p. 11.
76 *Hansard*, CXVIII, 1851, p. 12; *Select Committee on Patents*, Parl. Papers, XI, 1871, p. 674.

77 *Select Committee on Patents*, Parl. Papers, XXIX, 1864, p. 117.
78 *Mechanics Magazine*, LV, 1855, p. 32; I. Brunel, *The Life of Isambard Kingdom Brunel*, 1870. He talks of those 'abominable patent days', pp. 450–3, 454, 484, 489, 497.
79 *Select Committee on Patents*, Parl. Papers, XVIII, 1851, p. 271.
80 *Journal of the Royal Society of Arts*, 1852–53, pp. 271, 480–2, 525–7.
81 For an interesting discussion of the mid-nineteenth century debate see, V. M. Batzel, 'Legal monopoly in liberal England: the patent controversy in the mid-nineteenth century', *Business History*, 1980, pp. 189–202.
82 M. Davenport-Hill to Mr Grove, Q.C. 10.6.1867, cf. R. and F. Davenport-Hill, *A Memoir of Matthew Davenport-Hill*, 1878, pp. 148–9.

# 2

# The objectives of patent reform and the emergence of the invention interest

The patent system was persistently criticised during the industrial revolution, not because it was based on principles which were considered unjust, but because it failed to fulfil the expectations of inventors and patentees. In the first half of the nineteenth century its failures and inefficiencies became more apparent and led to the emergence of a vociferous invention interest, which demanded extensive reforms. Although a few were prepared to see the system abolished altogether, the majority were concerned with improving it. Reformers did not envisage the creation of alternative methods of inducing and rewarding inventive activity: they approved of the patent system. In the event, it was not reformed until the Patent Law Amendment Act was passed in 1852. Minor alterations were made in 1835, 1839 and 1844, but throughout the industrial revolution it remained largely the same as that which operated in the seventeenth century. Reform was 'cautious and circumspect'. This, and the following, chapter will examine the forces which shaped the reform movement.[1]

Patent reformers were concerned with a wide variety of technical issues, and although they frequently differed on how the system should be changed, there was a consensus on three points. Firstly, the administrative machinery for granting patents was absurdly cumbersome and 'calculated ... to baffle and paralyse the efforts of a class so essential in maintaining the commercial pre-eminence'[2] of the country. Secondly, patents were expensive and an unnecessary tax on ingenuity. Finally, the law was ineffective in protecting the rights of inventors. With varying degrees of emphasis, these three themes dominated the demands of the invention interest. Significantly, they were all concerned with administrative change, not with the basic economic philosophy justifying patents.

The procedure for granting patents was laid down by the 1536 Clerks Act,[3] and before an inventor was granted a patent his application had to go through as many as ten offices. Petitions, warrants and Bills were prepared several times over, signed and countersigned, and gratuities paid at each office before the final patent was given the Great Seal.[4] The prospective patentee was, moreover, responsible for negotiating his petition through this tortuous labyrinth. During this lengthy process patents were also frequently opposed by the use of caveats. Usually a party interested in recent development within a particular field of the invention market (e.g. steam engines) would leave a caveat with the Patent Office and he would then be notified of any patent applications which corresponded with the invention briefly described in the caveat itself. These caveats, which were notoriously vague, enabled competitors to monitor and oppose almost any invention passing through the Patent Office. This frequently caused delays — sometimes up to six months — and, more seriously, increased the possibility of industrial espionage. It is not without some irony that the Great Seal Patent Office should have been located in what were the Bankruptcy and Lunacy Offices.[5]

The 1536 Act, which was passed primarily to ensure fees to unsalaried Crown officers in the Signet and Privy Seal Offices, also determined the cost of patenting. Between 1750 and 1852 patents could cost anything up to £400, depending upon the geographical extent of protection. Down to the 1852 Reform Act, separate patents had to be registered for England, Scotland and Ireland. The cost of an English patent varied from invention to invention, but was between £100 and £120.[6] Equivalent patents for Scotland and Ireland were equally expensive. John Farey estimated that a Scottish patent cost in the region of £100 and an Irish one about £125.[7] Not all patentees sought such comprehensive protection, and whilst few official statistics distinguish between English, Scottish and Irish patents, a survey made in 1849 for the period 1838–47 seems to support the view that only important inventions were fully protected. Predictably, most patents were taken out only for England: approximately 35% of the total number of patents were extended to cover Scotland, whilst protection in Ireland fluctuated between a range of 21% and 12%.[8] On average only 16% of the patents taken out between 1838 and 1847 were for all three kingdoms. On the not unreasonable assumption that these ratios were consistent over time, the total cost of patenting was appreciably lower than that claimed by many of the witnesses examined by both the 1829 and 1851 Select Committees. Many, though, still regarded £120 for an English patent as excessive.[9]

The problems associated with the specification — the description of the invention — and the burden placed on this document by the courts was the third and final theme taken up by the invention interest. This aspect of reform was essentially concerned with the value of protection, rather than the practice of granting patents.[10] The main problem here was that inventors were not permitted to amend the specification once it had been enrolled. This raised serious difficulties, especially in the late eighteenth and early nineteenth centuries, when judges set a number of patents aside because of some fault (often trivial) in the specification. To overcome this, many inventors suggested that patent applications should be examined before reaching the Great Seal stage. In the late eighteenth century the problems created by the specification were the major point of contention.

The demand for patent reform rarely emerged as an issue in the eighteenth century. There were few patentees and even fewer court cases. Nor was the cumbersome procedure of the Patent Office in any sense unique, at least by eighteenth-century standards. Manufacturing and inventing, moreover, remained an infant industry for much of this period and made little impact on the national economy. In the 1780s, though, economic circumstances had changed, and when Pitt introduced the fustian tax in 1784, to offset some of the imposing national debt incurred during the American wars, manufacturers began to organise themselves on a more or less permanent basis. Commercial committees were set up in a number of industrial towns, notably Manchester, Sheffield, Norwich and Glasgow, and when Pitt introduced his free-trade Irish proposal in 1785, this led to the creation of the General Chamber of Manufacturers. In the 1780s British manufacturers reached a peak of organisation, and 'lobbying of government grew to national proportions'.[11] Significantly, this period also saw the beginnings of a move to reform the patent laws.

In June 1785 a number of patentees gathered in London to discuss Pitt's proposed Irish Commercial Treaty.[12] On 6 June they presented a petition opposing several clauses in the Bill which appeared to allow inventions patented prior to 1785 to be used and made in Ireland and then freely imported and resold in England. For James Watt, who gave evidence before a committee of the House of Lords on 14 June,[13] this was likely to invalidate 'the whole patent rights of this country, and thereby to do an act of the greatest injustice to a few individuals whose labours deserved better treatment'.[14] Many manufacturers were also opposed to the Irish Bill, but for quite different reasons.[15]

# The objectives of patent reform 37

In the same month an anonymous letter was circulated to a number of inventors urging the formation of a Patentees' Association to 'unite in defence of their respective rights and to agree upon a mode of application to Parliament for the better security of their inventions'.[16] The reason for the formation of a Patentees' Association was the complaint that a 'vast number of opulent manufacturers have agreed to use very beneficial patent inventions and have subscribed large sums to attack the same by writ of *Scire facias*'. Although the circular did not mention any particular case, it was clearly prompted by the threat to Arkwright's patent. Throughout the early 1780s Arkwright's patent had been persistently opposed by cotton manufacturers, and in 1785 Robert Peel organised a campaign through the Committee for the Protection and Encouragement of Trade, which was instrumental in bringing the patent down. Lancashire had broken 'through the privileges of the early patentees',[17] and for Manchester this was, as the *Derby Mercury* noted, 'almost of as much importance ... as the repeal of the Fustian Tax'.[18] For inventors, however, it was a serious blow, and they now felt that they would soon experience similar treatment. 'I have no doubt,' Watt observed, 'but we shall next be set up as a mark to be shot at and ruined if possible.'[19] Whether the patentees in London associated the cancellation of Arkwright's patent with the proposals included in the Irish Bill is unknown. Watt was certainly intrigued by the idea and suggested that the dismissal of Arkwright's patent was a ruse of the government to appease the Manchester cotton men's opposition to Pitt's treaty:

> Perhaps Mr ... A[rkwright]'s cause was determined before it came into court, and by the same kind of law and testimony any patent may be overthrown. I had a suspicion at the time that A[rkwright] was given up as a sugar plum by the M[iniste]r to the men of M[a]n[cheste]r to slacken their opposition to the Irish proposition ... you see how much we are in the power of these rascals.[20]

In the event a meeting was arranged, and on 25 June several patentees met at the Crown and Moll tavern in Chancery Lane. Watt was invited but did not attend, because, as he later wrote, 'at the last meeting of patentees [which was presumably the one before the committee of the House of Lords] I saw much a motely [sic] crew of projectors and madmen, some of which I thought it a disgrace to keep company ... I would far less associate with them.' Moreover, they 'managed their matters so ill that if they do not better ... they will get nothing but disgrace'. Watt claimed that the 'only decent persons that I recollect was Mr. Maurice, one Else from Nottingham and Marsh the stocking weaver but he talks eternally and not always sense'.[21] Arthur Else was the only inventor of the three and his interest was no doubt amplified by the knowledge that a writ of *Scire facias* was being brought against his only patent.[22]

Watt had other reservations. Public meetings, he reckoned, merely served to put 'the enemy on their guard' and gave them time to deploy 'their greater wealth' in forestalling or defeating any petition which might be presented to Parliament. His application for a patent extension in 1775 had taught him this much. 'Any combination of patentees to support one another would,' he claimed, 'be irregular, and ... however willing we may be to espouse the interest of any man of ingenuity we cannot think of making ourselves obnoxious to the public by supporting patentees whether they are in the right or in the wrong.' Instead he proposed to seek patent reform through 'using interest with persons in power'.[23] Anonymous lobbying by a few respectable patentees was likely to achieve more than noisy agitation.[24]

Watt's views on the proposed Patentees' Association were further influenced by his partner's business activities. In 1784 Boulton had become financially involved in Argand's patent oil lamp and, in return for helping with the cost of patenting, had been granted the exclusive rights of manufacturing. The patent promised to be a profitable investment, but the ease with which the lamp could be made encouraged a flood of infringements.[25] Throughout 1785 legal action was considered and, since Watt was likely to be an important witness, he refused to associate with any public demand for reform because he felt it would prejudice his own position.[26] Yet he did not completely ignore the Patentees' Association. He authorised Matthews to attend meetings and in this way also influenced developments.

Nothing appears to have come of the June meeting, but in September Matthews sent Watt a memorial on patents passed by the Committee of Patentees, as it was then called. This had been written by Abraham Weston, Watt's solicitor, and from its tone it would seem that Watt, or Matthews acting on his behalf, had more than a passing influence.[27] Watt certainly thought it 'very sensible and perfectly consistent' with his own 'sentiments on the subject'.[28] The memorial, however, was not a revealing document. It merely outlined a history of patents and showed that patentees were not sure what the law was regarding the specification. Weston hints that the committee wanted the administration of the law to be taken 'out of the hands of the ordinary courts of justice', but warned that this would be 'unpopular and strongly opposed'. He concluded the memorial with a comment which reflected the confused state of affairs: 'with regards to any particular plan for altering the law I am much at a loss, having been unable to think of any myself'. Later in the year Weston sent the memorial to Lord Liverpool under the misleading title 'Heads of a Bill for Explaining and Amending Laws Relative to Patents'.[29] Since no reforms were suggested the matter was ignored.

The failure of Arkwright's patent and Pitt's proposals still alarmed Watt. In July he wrote to Wedgwood, to see what could be done. His disapproval of recent events was plain:

> I agree that by the late decisions we have seen to what lengths the arm of despotic law may be stretched to undo any man who is suspected of the heinous crime of getting rich by his ingenuity. The same spirit of levelling prevailed there that has shown itself in forming the Irish resolution. The one was taking away the exclusive right of a private person to the productions of his own ingenuity and the other was the wholesale taking away the exclusive privilege of the nation.[30]

In this month Watt began to prepare his own proposals for reform.[31] He clearly thought deeply on the subject, as several versions remain extant.[32] Although the draft proposals vary in detail, they all emphasised the unnecessary burden which the courts placed on the specification, the need for improving the administrative procedure, and for redefining the statutory definition of what constituted a patentable invention. Watt proposed three significant changes to improve the security of the inventors: firstly, inventors should draw up two specifications rather than one; secondly, these should not be disclosed to the public during the life of the patent; and finally, all specifications should be examined by a Commission.

Prior to 1734 patents were granted on the condition that a number of tradesmen were taught the mysteries of the art. After the 1778 Liardet case this was to be done by the specification. Watt claimed that the change was grossly unwarranted and exposed inventors not only to cheating but to the whims of the judges. He therefore sought the reintroduction of the old form of disclosure during the patent's existence. Only after it had lapsed would the specification be open to the public. He also proposed changing the nature of the specification itself and suggested that, in the first place, it should simply describe the principle of what was new. A second specification would then be enrolled in Chancery a year later describing one way of putting principles into practice. Moreover, he wanted all improvements to be entered in a special improvement specification (which would expire with the original patent), to avoid the cost of taking out a quite separate patent.

Watt's final proposal concerned the examination of specification by a commission. The intention here was to eliminate the chance of patents being set aside because of some error in the specification itself. He suggested that the commission should include three members of the Royal Society, recommended by the Society's council, and two 'eminent artisans skilled in the art' relating to the patent, chosen by the Master of the Rolls, from a list of ten candidates provided by the patentee. The commission's powers were to be limited: they were 'not to be empowered to judge the merit of the invention, the novelty or utility thereof, but simply whether or not the patentee specified the same clearly or intelligently, but if they shall judge the same not to be a new invention, they are required to warn the patentee'.[33] Moreover, once the invention had received the commission's approval the courts could not set the patent aside on any grounds relating to the specification. The courts, it seems, were able to decide only on matters of novelty.

How far Watt's 'Thoughts' were supported in detail by other inventors is difficult to say. Arkwright read and commented on the proposal but appears to have added little of substance,[34] despite Wedgwood's hope that 'you two great genius's [sic] may probably strike out some new lights together which neither of you might think of separately'.[35] Arkwright, in fact, was not too happy about concealing the first and general specification, although he was prepared to hold back the second and more detailed specification until the patent expired. This probably reflected what most patentees would have liked.[36]

Watt does not appear to have circulated his ideas immediately. The dismissal of Arkwright's and Argand's patents together with the failure of the Patentees' Committee probably convinced him of the need to move cautiously. In 1785 and 1786 the atmosphere was not propitious for advancing radical suggestions for reform. As it turns out, nothing more was done until the early 1790s, when he himself considered legal action. Copies were then sent to politicians and other persons in power, and Boulton, as he had in the past, sought to exploit his own contacts.[37] In May 1792 Boulton wrote to the Attorney General, Sir Archibald MacDonald:

> ... We find ourselves urged both by the injuries we have sustained and by the insults of our opponents to seek redress by an action at law ... In this emergency we presume to look up to you Sir as one who is able and we hope not unwilling to procure us timely protection. The protection we have in our eye as least partial and most useful to other men of ingenuity is a

general law to explain and amend the law relative to patents, which in our opinion can with more propriety be moved for by you than by any other person ... If you think an alteration of the Patent Laws is necessary and proper and that our difficulties, as well as those of other patentees, form any reason for accelerating that event, we shall consider your patronizing it as a very great favour done to us as well as the public [sic] and shall be ready to give every assistance ....[38]

Boulton advised MacDonald that a copy of Watt's proposals could be got from Abraham Weston.

It is impossible to assess how many persons in power read the 'Thoughts'. Lord Kenyon probably received a copy in August 1790,[39] and J. Anstruther seems to have read a version in either late 1792 or early 1793. A Mr Bernard returned a copy to Anstruther in June and said that he was now 'more than ever convinced of the necessity for altering the present system on that head'. Mr Serjeant Watson also asked for a copy shortly after having spoken to Boulton and Watt some time in early or mid-1793.[40] How many of these had read the 'Thoughts' before May is unclear, but early in that month Mr Bernard brought a Bill before the Commons with the intention of concealing specifications (especially from foreigners) for the duration of a patent.[41] This was opposed in the Lords by Stanhope and Radnor and ultimately dismissed.

The failure of the Bill was probably expected. It involved a radical alteration in the argument which justified the granting of patents. Pressure of sorts, though, resurfaced for a while in 1795. In April Edmund Cartwright wrote to Watt urging the petitioning of Parliament, and was told that Mr Erskine 'had some plan ... in view but we are ignorant of whether or not he means to bring it forward during the present session of Parliament'.[42] Two months later a 'Call to Patentees' was published:

> It appearing to some persons interested in Patents and who may have occasion to apply for others, that some further regulations are wanted as to the making and entering of caveats and specifications for the better security of their discoveries, in making trials before they obtain patents and in securing the rights granted by them, they are desirous of meeting with those who have the like interest, in order to their considering the same, and propose enacting one, at which time and place they request the favour of the Company of Patentees or of their agents.[43]

What resulted from this advertisement is unknown, but it seems to have marked the end of the demand for patent reform in the eighteenth century.

## 42  The patent institution

Pressure for reform lapsed during the Napoleonic wars, but in 1820 and 1821 two Bills were brought before Parliament by several inventors, engineers and manufacturers from London. A number of reforms were proposed: protection during experimentation; a body of commissioners consisting of engineers and scientists; the amendment of small errors in the specification; and a published list of existing patents. Significantly, neither Bill suggested reducing the cost of patenting, and both failed to pass beyond a second reading,[44] largely because Parliament felt that inventors were attempting to alter the social contract between themselves and the public. Since the 1790s almost all inventors who petitioned Parliament wanted to keep their specifications secret, on the grounds that this would prevent imitation by the French. James Booth in 1792, and James Lee in 1812, both succeeded in having their specifications concealed for seven years.[45] Other applications met with uncompromising resistance, as did the 1820 Bill brought in by Wrottersley and Gilbert to prevent the general procurement of specifications.[46] Although the two other Bills did not even suggest concealing specifications, the ill-timed introduction of the Wrottesley proposal must have created serious doubts about any reform of the system.

In the middle of 1825, as trading conditions showed signs of deterioration, the desire for patent reform slowly re-emerged. In June Joseph Crowder, inventor of a bobbin net machine, attempted to interest Joseph Hume in the possibility of reform, but Hume declined to help because he thought the patent laws were far too complicated.[47] In the same month Gravenor Henson sought the support of Francis Place. His strident claim that the 'privilege of granting patents has been intrusted to the most incompetent and improper persons that can possibly be selected among the servants of the Crown' drew no response.[48] A month before the financial crash of December, the Manchester Chamber of Commerce set up a committee to prepare a paper on the patent laws. Patents, it was agreed, 'had become desirable, most especially at a period when the manufacturers of this country have to sustain an active competition with the production of foreign industries'. Since prosperity depended upon the 'superiority of our inventions for facilitating production the [patent laws] should be made as perfect as its nature will admit'. The committee's proposals were similar to the 1820 and 1821 Bills, but no one was prepared to take the initiative, preferring instead to wait and see what the Attorney General intended doing.[49] He agreed that patent reform was a subject of 'the greatest importance' but insisted that changes in the system

# The objectives of patent reform

had to be 'consistent with policy and sound reason',[50] which in effect meant doing nothing.

In the following year the *Mechanics Magazine* openly committed itself to patent reform. In one of many editorials Joseph Clinton Robertson, a practising patent agent, stressed the importance of the press: 'seeing now no other prospect of the emancipation of mechanical genius except through the immediate agency of the press, we propose to insert forthwith such of the numerous communications we have received on the subject, as appear to us to bear most forcibly upon it'.[51] Several anonymous letters urged inventors to unite, and one writer concluded that 'it will be entirely the fault of the mechanics themselves if the present laws are not altered'.[52] A public meeting was soon called at the Mechanics Institute, Chancery Lane, but it turned out to be a miserable affair. An hour after the meeting was to have started only half a dozen had arrived. Dr Birkbeck, Mr Rotch and Major Shaw were three named by *The Times*, and Shaw was forced to take the chair, as the proposed chairman, the Marquess of Clanricarde, had absented himself.[53]

The failure of this meeting makes it difficult to explain why, two years later, the government appointed a Select Committee on patents. The *Mechanics Magazine* immediately congratulated its readers for having brought the matter to the attention of Parliament, but provided no evidence to support this rather grand claim.[54] No articles or letters on patents had appeared in the *Magazine* since the Chancery Lane fiasco. Thomas Webster, in his short account of patent reform, suggests that several reform committees were set up in 1828 but mentions no details, except about the one formed in Manchester. None of the usually vociferous chambers of commerce appears to have petitioned Parliament or lobbied individual members of the House at this time. In fact, only two petitions were presented to Parliament, one by John Birkinshaw and the other by an anonymous group of manufacturers, engineers and inventors,[55] but both were presented after Parliament had agreed to the setting up of the committee. Lennard, who introduced the motion, was himself an ardent supporter of the patent system and, according to another anonymous letter in *The Times*, had for a number of years wanted a committee of inquiry.[56] But why he should seek an inquiry in 1829 remains unknown.

The motion for an inquiry into the patent laws was brought before the House by Lennard on 9 April, and was seconded by Mr Davies Gilbert. Both emphasised the difficult nature of the subject and sought to make a virtue of caution. Lennard assured Parliament that he had no intention of impugning the principles on which the law was founded or of encroaching on the royal prerogative. Robert Peel endorsed the need for circumspection, but opposed any reduction in the cost of patents because, he feared, industrialists in places like Birmingham and Manchester would be constantly locked in vexatious legal action over the consequent increase in the number of trivial inventions. Lennard was less willing to commit himself on this important issue, but appears to have agreed with Peel.[57] As far as the *Mechanics Magazine* was concerned there were too many 'busy in the field to whom the interest of the poor man and the humble are of no moment'. Robertson genuinely believed that the opinions expressed in the Commons did not 'warrant the hope of much good being done', and feared that the witnesses invited to give evidence would have 'a superficial and crude knowledge of the deficiencies of the law'.[58]

Some of Robertson's fears were justified by events. Of the nineteen witnesses examined by the committee, only six were directly involved with manufacturing and inventing. The remaining thirteen were drawn from a variety of occupations professionally associated with patents: engineers, solicitors, journalists, patent agents and Patent Office officials. There were no representatives from the cotton, chemical, machine-making or iron industries. Joseph Merry, a Coventry ribbon manufacturer, was the only witness from the textile industry, and he, like other witnesses, displayed a rather crude knowledge of the system as a whole. John Taylor, who manufactured gas from oil, admitted that he had not thought much on patents, and Marc Brunel claimed he knew very little about the subject, despite owning a number of patents himself.[59] Apart from John Farey, who provided the bulk of the evidence, there was no overall assessment of the deficiencies of the system and their effect on inventors generally. The 1829 committee was not, on the other hand, a committee of convenience and, although in the 1830s the manipulation and management of select committees were common practice,[60] there is no evidence to suggest that any of the witnesses in 1829 were invited simply because they were of the 'same mind as those who invited them'.[61]

Four problems repeatedly emerged from the evidence: patents did not provide adequate protection because the law was uncertain and because the delay in granting patents allowed competitors time for prior publication; the burden placed upon the specification was far too great; the fixed term for patents precluded some inventors from reaping rewards for inventions which had a long pay-off period; and, finally, patents were too expensive. The establishment of a scientific commission to examine inventions was the main proposal and was recommended by a number of witnesses.[62] Such a commission, it was believed, would reduce the number of faulty specifications and thereby the risk of patents being set aside by the courts. The commission would also permit patents to be sealed at the time of application, and allow inventors to alter the specification along the lines suggested by the examiners, as happened in France. Predictably, Patent Office officials were reluctant to see such changes. Francis Abbot, who was employed in the Petty Bag Office, claimed that it would be difficult to select examiners and to avoid the problem of partiality.[63] Moses Poole, a clerk in the Patent Office, and a practising patent agent, thought such matters were best left to the courts.

Several witnesses wanted the cost of patenting reduced,[64] though none argued the case with conviction. Significantly, Farey (who certainly impressed the committee with his knowledge of patents) and Brunel opposed cost changes on excactly the same grounds that Peel had urged in Parliament: cheap patents would lead to an increase in the number of trivial inventions and, in turn, to an increase in the number of costly lawsuits.[65] William Newton, another practising patent agent, suggested that costs might vary with the length of a patent, and recommended shorter patents for inventions which proved to be less important.[66] Taylor, Farey, Brunel and Clegg were all in favour of varying the length of a patent and generally thought that fourteen years was insufficient for a large number of inventions.[67] In short, on the issue of costs, there was little unanimity.

Writing in 1852, Thomas Webster suggested that the witnesses in 1829 were 'almost unanimous in condemning the existing system and were agreed on many material points as to the remedy to be applied'.[68] J. R. McCulloch was less enthusiastic, but agreed that the evidence consisted of 'a great deal of curious and instructive information'.[69] A century later Harding took a much more critical view: 'the comparatively mild criticism of the extraordinary series of steps necessary to obtain a patent [indicates] how strongly established the practice had become'.[70] These various assessments are, in a sense, all accurate. The committee did produce a mass of interesting, as well as anecdotal, information; there was wide criticism of the existing system but individual witnesses were not forceful enough; several remedies were suggested but never with much commitment. In fact the witnesses seem to have been every bit as cautious as those who moved to set up the inquiry in the first place. In the event the committee did not report.

The inconclusive findings of the committee and its failure to send in a report did not dampen the enthusiasm for reform to the extent predicted by the *Mechanics Magazine*,[71] although there is very little doubt that many must have been disappointed to learn that the committee was not going to be reappointed. In 1830 Richard Roberts published a detailed set of proposals for changing the law,[72] but in that year the demand for parliamentary reform was fast becoming the most pressing national issue and eclipsed all hope of anything further being done in the immediate future.[73] After 1832, though, the demand for patent reform quickly re-emerged. Meetings were held in Birmingham, Manchester and Leeds[74] and in 1833 another Bill was introduced.

The 1833 Bill was brought before Parliament on 19 February by Godson, Lennard and Phillpots.[75] Many of its proposals reiterated what had been sought in 1820 and 1821, except that the idea of establishing a Board of Scientific Examiners had now been abandoned. The Bill was also rather cautious about reducing the cost of patenting, and proposed that patents should be granted for a period of seven or fourteen years; shorter ones were to cost a third of a fourteen-year patent, the cost of which was to remain unchanged. The Bill also proposed that patentees could enter a provisional specification which could then be amended by a complete specification within a year, so long as this did not materially affect the meaning of the original specification. Judges, it was further proposed, should also have the power to amend the specification 'in all matters of form and description'. Patents, moreover, should not be made void in consequence 'of the same being imperfectly or privately used previously', or if 'not practically used in a public manner within the last ten years'.[76]

The Bill, which was written jointly by three solicitors, Richard Godson, Benjamin Rotch and Archibald Rosser, met with opposition both in and outside Parliament. Some reformers were severe in their criticism. The *Mechanics Magazine* was never happy, and thought the Bill 'excessively careless and slovenly [and] ... faulty'. Mr Godson, it argued 'has evidently undertaken a task for which he is unequal', suggesting that 'the sooner he abandons it the better'.[77] The Birmingham Chamber of Commerce was no more sympathetic. 'The Godson bill,' the minutes noted, 'is defective in most of its provisions and if passed into law would fail to secure to inventors that protection and facility which it was the object of patent laws to do.'[78] The Manchester Law Society was equally hostile, and very aggrieved that Godson had ignored the views sent to him by a 'considerable body of engineers and mechanics in Manchester'.[79] Richard Roberts went so far as to publish a lengthy amendment to the Bill, a copy of which he sent to Lord Brougham, who himself thought the Bill rather 'faulty'.[80] According to Alexander Kay its defects were so apparent there was no need for debate.[81]

It seems unlikely that Godson intended his Bill to do what others thought it would do. Yet there were several clauses which were far from satisfactory. The suggestion that any person to whom an invention had been communicated ought to be able to take out a patent was ill-conceived. John Farey observed that it exposed inventors to all kinds of cheating, especially during the experimental stage, and encouraged competitors, who frequently used industrial spies and bribery to keep abreast of new technical developments, to patent the ideas of others without fear of being taken to law.[82] Moreover, the notion that the buyers of an invention should also be able to take a patent out in their own name raised similar difficulties. For inventors who sold rather than used their inventions patents provided the only form of security, and as a rule buyers were reluctant to take up inventions without this protection. Patents were also an extremely useful marketing device, and if inventors were encouraged to surrender their name rights it would have made it increasingly difficult to establish their inventive pedigree, and would certainly have diminished their future bargaining power with the buyers of inventive output. 'Had the authors of the bill,' the *Mechanics Magazine* acidly noted, 'put their heads together to contrive how they might best expose inventors to be circumvented and cheated they could not have framed a clause better calculated for the purpose.'[83]

The clause which attempted to redefine what could be the subject of a patent was equally ill-conceived. This part of the Bill was prepared by Benjamin Rotch. He proposed that patents could be taken out for 'all new substances or things made ... all new machines ... all new commodities or arrangements or machinery or things, either already known or discovered ... all principles, new discoveries and all new applications, [and] all chemical discoveries'. Many considered this too wide, since it allowed almost anything to be patented, including things already known. Robertson preferred the old Statute of Monopolies,[84] and so did many others.[85] Even Archibald Rosser thought the clause unsatisfactory and tried to persuade Rotch to leave it out.[86] Godson later sought to divide the Bill into two parts, one concerned with administration and the other with the cost of patenting,[87] but there was insufficient time and the matter was left over until the following session.[88]

The failure of the 1833 Bill is revealing in a number of respects. Firstly, it showed that there was no unambiguously agreed remedy for rectifying the faults of the system. Although most reformers, acting individually or in organised groups, were anxious to improve protection for inventors, there was little agreement about the means of bringing it about. Reform groups generally acted without reference to the proposals advanced elsewhere, and their disagreement with the Bill appears to have been the major ground for agreement. Reformers were vociferous but not united.[89] Secondly, the authors of the Bill made several tactical errors and appear to have been influenced by previous arguments concerning cheaper patents and the establishment of a Board of Examiners. Rosser's claim that imperfections were bound to arise when attempting to legislate on such a difficult subject[90] seemed rather limp in the circumstances and was, no doubt, treated with contempt by those whose advice he had ignored. Finally, although the government and Patent Office officials seemed willing to accept some measure of reform, the hostile reaction to the Bill allowed them to postpone the matter without giving the impression that there were some who actively opposed sweeping changes.[91] Significantly, the promise that reform would be one of the first issues to be considered in the next parliamentary session was never fulfilled, despite further petitions.[92]

The most important outcome of the 1833 debate was the growing interest of Lord Brougham in patent reform. In the 1820s and early 1830s several individuals had unsuccessfully urged him to take up the issue.[93] But during the passage of the 1833 Bill, Rosser appears to have persuaded him to consider what might be done.[94] The large number of letters he received must also have impressed upon Brougham the need for reform, and in 1835 he introduced his own Bill.

Compared with Godson's, it was rather unambitious. The main proposal was to use the Judicial Committee of the Privy Council to investigate whether particular patents should be extended. Normally, this could be done only by expensive private Act of Parliament. Brougham also suggested that the committee should serve as a board of examiners over disputes concerning inventive priority. Finally, he proposed that patentees could alter parts of the specification by a disclaimer.[95] Rosser thought the Bill 'would be an excellent beginning', but was anxious that legislation 'not carried through [would] only serve to [render] the prospects of amendment remote'.[96]

This anxiety was well founded. A petition from patentees, merchants, solicitors and manufacturers of Birmingham opposed the Bill because it failed to consider reducing the cost of patenting. A further amendment was proposed by James March of London,[97] and the *Mechanics Magazine* weighed in with its usual hyperbole. The Bill 'is a very crude and ill-digested affair ... His lordship is evidently but very imperfectly acquainted with the actual defects of the law ... he does not know where the shoe pinches, though truth to speak, the whereabouts is no great secret.'[98] Several MPs argued that the Bill did not go far enough to remedy the 'evils of the present system', and Mr Mackinnon was astonished that such a man as Lord Brougham 'could countenance and bring forth such a miserable bungling piece of legislation'.[99] The Bill was eventually referred to a select committee which met behind closed doors.[100]

Only three witnesses were called, and all — without names being mentioned — were savagely lampooned by the *Mechanics Magazine*. Firstly, there was 'the solicitor to the rejected bill of 1833 who, saving his share in that abortion, had no call to bear witness on the subject, having confessedly no practical experience of the evils to be redressed'. Secondly, there was 'a gentleman who had raised himself by his inventions and patents from a very humble condition in life to affluence and distinction but who for the very reason that he has been so fortunate is but an indifferent judge of the difficulties which the generality of poor inventors have to encounter'. Lastly, there was 'a consulting engineer of considerable standing who went prepared to occupy the committee for a couple of days with his crotchets and speculations, but who was civilly dismissed after a few questions asked and answered and who, it must be confessed, was likely to have been more verbose and tedious than either pertinent or edifying'.[101] The three witnesses were Alexander Rosser, John Heathcoat and John Farey.

The *Magazine*'s parody was excessive, although not entirely unwarranted. Rosser, who admitted having little practical experience of patents, was extremely cautious and seemed content to improve the position of inventors by small steps. He believed it was wrong to leave patentees unprotected any longer, but warned against the danger of attempting to clog the Bill with too many proposals. Experience had shown that it was counterproductive. Heathcoat was equally cautious. He noted that the alleged defects in the specification provided the greatest difficulty for inventors, and claimed that many were forced to compromise under the collusive pressure of infringers. On other matters he was silent, and informed the committee that he was unwilling to 'think aloud his own ideas' where they fell outside his brief. On the evidence it is difficult to accept Rosser's unconditional judgement that Heathcoat was well 'versed in the law of patents'.[102] The *Magazine*'s attack on John Farey was, however, totally unwarranted. His evidence (as in 1829) was informed by practical experience and by obvious reflection. And, except for one or two doubts concerning the administration of the law by the courts,[103] he approved of the Bill in every respect and predicted it would 'produce most important benefits to the nation'.[104]

Others were less convinced, and in Parliament the Bill experienced further difficulties. Tooke, Parks, Lennard, Bowring, Prime and Mackinnon believed it to be inadequate but, as it was too late in the session to bring in a more comprehensive measure, they were prepared to see it as the foundation for further legislation. Some remained utterly opposed to granting the Judicial Committee of the Privy Council the right to prolong patents. For Brougham this was a serious threat. He believed it was the best part of the Bill, which he was prepared to withdraw altogether if the clause was removed.[105] In the event, with the help of Rosser who went 'round canvassing every member for support',[106] he managed to win a majority of two and, on 10 September, the Bill was given the royal assent.[107]

In so far as the 1835 Act was the first piece of legislation on patents since the Statute of Monopolies it was a victory for inventors. The victory, though, was limited in scope. Patentees could now amend their specifications to remove small errors[108] and petition the Judicial Committee of the Privy Council to extend their patents for a further seven years, but the system was left very much as it was. The reaction of inventors and those actively involved in reform is difficult to gauge. Richard Godson praised Brougham's great skill in carrying the Bill through and considered it 'a great boon to inventors'.[109] William Newton was less enthusiastic but recognised the Act as an important concession.[110] The *Mechanics Magazine* predictably remained unconvinced. It approved of the Judicial Committee of the Privy Council, but genuinely believed that the Act did nothing to resolve the real problem: 'On every former occasion when the Patent laws have been under the consideration of Parliament, the dangers of making patents cheap has been a favourite theme; but no one seems disposed to risk his character for commonsense by talking in this way now.'[111]

The absence of evidence on how inventors felt suggests that some were quite happy to see some progress being made. Lord Brougham's initiative may not have done all they wished, but his intervention at least held out the promise of further change. Brougham was, after all, the leading legal reformer in the 1830s, and many probably realised that his continued support justified a measure of compromise. Reform by instalment was better than none at all.[112]

Brougham's Act did not, however, satisfy all the critics,[113] and in the following year a number of petitions from London and the West Country urged further alteration, especially in the cost of patenting.[114] W. A. Mackinnon, who had been Brougham's severest critic, presented another comprehensive Bill reiterating all the demands made in 1820 and 1821, as well as reducing patent fees. It was supported by several more petitions from patentees, inventors and manufacturers, but did not proceed.[115] In 1837 Mackinnon introduced another Bill with very few modifications.[116] This again received fairly wide support.[117] Gravenor Henson, chairman of the Society of Inventors, considered it 'would do us more good than any other thing we know'.[118] The *Mechanics Magazine* suggested a few changes but believed, quite uncharacteristically, that the Bill reflected what inventors wanted.[119] In Parliament several paid lip service to the need for reform, but when the Attorney General said he could not bring himself to accept such sweeping changes the Bill was effectively prevented from going any further.[120] In the late 1830s patent reform still had a long way to go.

**Notes**

1 Boehm and Silberston have already given a short account of the patent reform movement, but were unable to tell the story in any detail: *op. cit.*, pp. 26–9. See also Macleod, *op. cit.*, pp. 25–53, 278–80.
2 *Sessional Papers of the House of Lords*, XVI, 1851, pp. 456–8.
3 27 Henry VIII, cap. 11, s. 8; W. M. Hindmarch, *Patent Laws of this Country: Suggestions for the Reform of them*, 1851, pp. 3–5; S. E. Lehmberg, *The Reformation Parliament, 1529–1536*, Cambridge, 1970, p. 239.
4 See Gomme, *op. cit.*, 1946, and C. Dickens, *A Poor Man's Tale of a Patent*, 1925 ed.
5 A. A. Gomme, 'Centenary of the Patent Office', *Transactions of the Newcomen Society*, 1951–53, p. 165.

## 52  The patent institution

6 Diary of Samuel Taylor, MS/925/775. Manchester Reference Library (hereafter M.R.L.).
7 *Select Committee on Patents*, Parl. Papers, III, 1829, p. 17.
8 *Select Committee on Patents*, Parl. Papers, XVIII, 1851, p. 40; *Hansard*, CXVIII, 1851, p. 1920; *Select Committee on Patents*, Parl. Papers, XVIII, 1851, p. 429. As it was exceedingly rare for inventors to hold a Scottish or an Irish patent without an English one, the number of English patents can be regarded as the annual total of patents registered.
9 *Parl. Papers* XLV, 1849, p. 383.
10 *Select Committee on the Design Act Extension Bill*, Parl. Papers, XVIII, 1851, p. 695.
11 J. M. Norris, 'Samuel Garbett and the early development of industrial lobbying in Great Britain', *Economic History Review*, 1957–58, pp. 450, 460; M. Kammen, *Empire and Interest: the American Colonies and the Politics of Mercantilism*, Philadelphia, 1970, p. 139; T. S. Ashton, *Iron and Steel in the Industrial Revolution*, Manchester, 1963, 3rd ed., p. 164.
12 W. Bowden, 'The influence of the manufacturers on some of the early policies of William Pitt', *American Historical Review*, 1924, p. 688.
13 Boulton and Watt MSS, Boulton to Watt, 10.6.1785, Parcel E, Birmingham Reference Library (hereafter B.R.L.).
14 Boulton and Watt MSS, Watt to Lord Loughborough, 8.6.1785, B.R.L.
15 See J. Holland Rose, *William Pitt and the National Revival*, 1911; V. T. Harlow, *The Founding of the Second British Empire, 1763–1793*, I, 1952; J. A. Langford, *A Century of Birmingham Life, 1741–1841*, Birmingham, 1868, pp. 320–9.
16 Boulton and Watt MSS, Watt to Matthews, 20.7.1785, Letter Book (Office Steam Engine), B.R.L.
17 J. Wheeler, *Manchester: its Political, Social and Commercial History*, Manchester, 1836, p. 160.
18 *Derby Mercury*, 23–30.6.1785.
19 Boulton and Watt MSS, Watt to Matthews, 20.7.1785, Letter Book (Office Steam Engine), B.R.L.
20 *Ibid.*
21 *Ibid.*
22 *Rex* v. *Else* (King's Bench) 1785; G. Henson, *History of the Framework Knitters*, Newton Abbot, 1970 ed., pp. 315–16. Else had also been involved in patent litigation in the 1760s. *Select Committee on Patents*, Parl. Papers, III, 1829, p. 186. Else's patent was for the manufacture of French or wire grand lace, N° 1235, 29.10.1779.
23 Boulton and Watt MSS, Watt to Matthews, 20.7.1785, Letter Book, B.R.L.
24 Boulton and Watt MSS, Watt to Wedgwood, 20.7.1785, Letter Book; Watt to Boulton, 21.7.1785, Letter Book; Watt to Matthews, 22.7.1785, 28.7.1785, Letter Book, B.R.L.
25 H. W. Dickinson, *Matthew Boulton*, Cambridge, 1937, pp. 128–9.
26 H. W. Dickinson, *James Watt: Craftsman and Engineer*, Cambridge, 1936, pp. 121, 133. *Argand* v. *Magellan*, 1786. This is one case not reported in Woodcroft's list.

## The objectives of patent reform

27  Boulton and Watt MSS, *Observations on Patents*, Miscellaneous Papers, Parcel E, 1785, B.R.L.
28  Boulton and Watt MSS, Watt to Matthews, 28.9.1785, Letter Book, B.R.L.
29  Add. MSS. 38,345, Liverpool Papers.
30  Boulton and Watt MSS, Watt to Wedgwood, 20.7.1785, B.R.L.
31  Boulton and Watt MSS, Watt to Boulton, 21.7.1785, Letter Book, B.R.L.
32  Robinson, *op. cit.*, 1970. Robinson shows that Watt had two versions of a paper entitled 'Thoughts upon Patents for exclusive Privileges for new Inventions', and two versions of another paper entitled 'Heads of a Bill to explain and amend the laws relative to Letters Patent and grants of privilege for New Inventions'.
33  Musson and Robinson, *op. cit.*, 1969, pp. 213–28.
34  Robinson, *op. cit.*, 1971, p. 129; E. Roll, *An Early Experiment in Industrial Organisation*, 1930, pp. 146, 284–6.
35  Boulton and Watt MSS, Wedgwood to Watt, 17.9.1785, Box 36, B.R.L.
36  In the 1790s almost all petitions to Parliament were concerned with the concealment of the specification. See below.
37  E. Robinson, 'Matthew Boulton and the art of parliamentary lobbying', *Historical Journal*, 1964, pp. 209–29; D. Bargar, 'Matthew Boulton and the Birmingham Petition of 1775', *William and Mary Quarterly*, 1956, pp. 26–39.
38  Boulton and Watt MSS, Boulton to MacDonald, 21.5.1792, Box 21, Bundle 3, B.R.L.
39  Musson and Robinson, *op. cit.*, 1969, p. 213.
40  Boulton and Watt MSS, Anstruther to Watt 27.5.1793, Parcel E; Bernard to Anstruther, 18.6.1793, Parcel E, B.R.L.
41  *J.H.C.* XLVIII, 1793, pp. 761, 766, 785, 807, 815, 821.
42  Boulton and Watt MSS, Watt to Cartwright, 28.3.1795, Letter Book, B.R.L.
43  Boulton and Watt MSS, Parcel E, B.R.L.
44  *Parl. Papers*, I, 1820, p. 277; *J.H.C.*, LXXV, 1820, pp. 338, 342, 362, 368, 399, 413; *Parl. Papers*, I, 1821, p. 21; *J.H.C.*, LXXVI, 1821, pp. 40, 51, 101.
45  *J.H.C.*, XLVII, 1792, pp. 499, 559, 1075, 1088, 1090; 32 George III, cap. 73; *J.H.C.*, LXVIII, 1812, pp. 84, 614–15; *L.J.*, CLXXIX, 1813, pp. 3237–40; 53 George III, cap. 179; *Parl. Papers*, I, 1820, pp. 258–9; *J.H.C.*, LXXV, 1820, pp. 312, 352, 357, 389, 402.
46  *Hansard*, I, 1820, pp. 1052–3.
47  Add. MSS, 27,807, Crowder to Hume, 6.6.1825.
48  Add. MSS, 27,807, Henson to Place, 7.6.1825. See also R. A. Church, *Economic and Social Change in a Midland Town: Victorian Nottingham, 1815–1900*, 1966, pp. 323–4; R. A. Church and S. D. Chapman, 'Gravenor Henson and the making of the English working class', in E. L. Jones and G. E. Mingay, *Land, Labour and Population in the Industrial Revolution*, 1967, p. 151.
49  Minutes of the Manchester Chamber of Commerce, M8/2/1, 1821–27, 9.11.1825, pp. 369–72, M.R.L.
50  *Hansard*, XV, 1826, pp. 70–6.

51 *Mechanics Magazine*, VII, 1827, p. 150.
52 *Mechanics Magazine*, VII, 1827, pp. 149–50, 270–1, 277–8, 313–16.
53 *Times*, 6.6.1827.
54 *Mechanics Magazine*, XI, 1829, p. 154.
55 *J.H.C.*, LXXXIV, 1829, p. 122.
56 *Times*, 8.6.1829.
57 *Hansard*, XXI, 1829, pp. 598–608.
58 *Mechanics Magazine*, XI, 1829, pp. 155–6.
59 *Select Committee on Patents*, Parl. Papers, III, 1829, pp. 7, 9, 38.
60 L. Brown, *The Board of Trade and the Free Trade Movement, 1830–42*, Oxford, 1958, pp. 5, 72–5; and also 'The Board of Trade and the tariff problem, 1840–42', *English Historical Review*, 1953, p. 402.
61 *Mechanics Magazine*, XI, 1829, pp. 155–6.
62 *Select Committee on Patents*, Parl. Papers, III, 1829, p. 55.
63 *Ibid.*, p. 55.
64 *Ibid.*, pp. 13, 72, 88.
65 *Ibid.*, p. 38.
66 *Ibid.*, p. 72.
67 *Ibid.*, pp. 10, 26, 38, 94.
68 Webster, *History ...*, 1852, p. 6.
69 J. R. McCulloch, *Dictionary of Commerce*, 1851, 3rd ed., p. 936.
70 Harding, *op. cit.*, p. 6.
71 *Mechanics Magazine*, XI, 1829. 'The present is such an opportunity as may not soon occur again ... and if lost may take the labours of many years to retrieve', p. 156.
72 R. Roberts, *Outlines of the Proposed Law of Patents for Mechanical Inventions*, Manchester, 1830.
73 *Patent Journal*, VIII–IX, 1848, p. 299.
74 *Newton's Journal*, XXXIX, Conjoined Series, 1849, p. 41.
75 *J.H.C.*, LXXXVIII, 1833, pp. 72, 130, 145, 297; *Hansard*, XV, 1833, pp. 974–88.
76 *Parl. Papers*, III, 1833, pp. 169–77, 183–7.
77 *Mechanics Magazine*, XIX, 1833, p. 48.
78 Wright, *op. cit.*, pp. 115–16.
79 Brougham MSS, [45,917], A. Kay to Brougham, 24.8.1833, U.C.L.
80 R. Roberts, *Outlines of a Bill to Amend the Law for granting Patents for Invention*, Manchester, 1833; see also *A Bill to Amend the Law Granting Patents for Invention: Suggested as an Amendment to Mr. Godson's Bill now before Parliament*, Brougham MSS, [43,814], [46,265], 23.11.1833 U.C.L. See also *Mechanics Magazine*, XXIII, 1835, pp. 171–2; *Hansard*, XXX, 1835, pp. 1186–88.
81 Brougham MSS, [45,917], Kay to Brougham, 24.8.1833, U.C.L.
82 Brougham MSS, [45,493], Farey to Brougham, 21.8.1833, U.C.L.
83 *Mechanics Magazine*, XIX, 1833, p. 297.
84 *Mechanics Magazine*, XIX, 1833, p. 228.
85 See, for example, Brougham MSS [8,754], W. Chubb to Brougham, 9.6.1835, U.C.L.; *J.H.C.*, LXXXVIII, 1833, p. 535.
86 Brougham MSS, [6,640], A. Rosser to Brougham, 15.7.1833, U.C.L.
87 *J.H.C.*, LXXXVIII, 1833, pp. 528, 559, 564, 565.

## The objectives of patent reform

88 *Hansard*, XIX, 1833, pp. 440–1.
89 William Newton writing in 1849 suggested that the Godson Bill failed because of a general apathy among inventors. *Newtons Journal*, XXXV, Conjoined Series, 1849, p. 43.
90 Brougham MSS, [6,640], A. Rosser to Brougham, 15.7.1833, U.C.L.
91 See, for example, *Newtons Journal*, XXXIV, Conjoined Series, 1849, p. 41.
92 *J.H.C.*, LXXXIX, 1834, pp. 117, 227. See also *Hansard*, XXX, 1835, p. 467.
93 Brougham MSS, [28,617], A. Galloway to Brougham, 1.3.1825; [30,824], A. Gordon to Brougham 19.12.1831; [30,723], C. Griffin to Brougham, 7.2.1831, U.C.L.
94 Brougham MSS, [6,640], A. Rosser to Brougham, 15.7.1833; [46,271], A. Rosser to Brougham, 8.11.1833; [46,272], A. Rosser to Brougham, 30.11.1833; [45,209], A. Rosser to D. Le Marchant, 26.7.1833 and [45,358], 1.8.1833, U.C.L.
95 *Parl. Papers*, III, 1835, p. 597.
96 Brougham MSS, [21,265], A. Rosser to Brougham, 11.6.1835, U.C.L.
97 *J.H.C.*, XC. 1835, pp. 578, 584.
98 *Mechanics Magazine*, XXIII, 1835, p. 17.
99 *Hansard*, XXX, 1835, pp. 466–7.
100 *J.H.C.*, XC, 1835, pp. 183, 217, 227, 231, 282.
101 *Mechanics Magazine*, XXIII, 1835, p. 310.
102 Brougham MSS, [21,269], A. Rosser to Brougham, 10.9.1835, U.C.L.
103 See Chapter 6.
104 *Minutes of Evidence taken before the Lords Select Committee on Patents Amendment Bill*, 6.7.1835, House of Lords Records Office (hereafter H.L.R.O.).
105 *Hansard*, XXX, 1835, pp. 466–71, 1186–88.
106 Brougham MSS, [21,269], A. Rosser to Brougham, 10.9.1835, U.C.L.
107 5 and 6 William IV, cap. 83; *J.H.C.*, XC, 1835, p. 661.
108 The disclaimer did not permit inventors to alter the meaning of the description in the specification.
109 R. Godson, *The Law of Patents for Invention and Copyright*, 1840, 2nd ed., pp. 21–2.
110 *Newtons Journal*, XXXIV, Conjoined Series, 1849, p. 41.
111 *Mechanics Magazine*, XXIII, 1835, p. 384. Throughout the passage of the Bill, Rosser had attempted to persuade Robertson of the need to accept reform by instalments. See *Mechanics Magazine*, XXIV, 1835, p. 31; *Times*, 5.10.1835.
112 Significantly, Broughton was not the only legal reformer who wished to see gradual change. Lord Lyndhurst also agreed that patent legislation should be confined 'to a very few topics'. Brougham MSS, [21,272], Rosser to Brougham, 8.10.1835, U.C.L.
113 Brougham MSS, [27,885], J. Woolams by Brougham, 5.7.1835, U.C.L.
114 *J.H.C.*, XCI, 1836, pp. 196, 416, 471, 750, 818.
115 *J.H.C.*, XCI, 1836, pp. 475, 526, 684, 709, 775, 823, 825; *Parl. Papers*, IV, 1836, pp. 513–21.

116 *Parl. Papers*, III, 1837, pp. 315–28.
117 *J.H.C.*, XCII, 1837, pp. 139, 152, 435; Brougham MSS, [23,061], R. R. Reinagle to Brougham, 25.2.1837, U.C.L.
118 *Select Committee on Postage*, Parl. Papers, XX, 1837–38, p. 230.
119 *Mechanics Magazine*, XXXV, 1837, pp. 487–91.
120 *Hansard*, XXXVI, 1837, pp. 554–8.

# 3

# The Patent Law Amendment Act 1852

During the early 1840s the demand for reform lapsed once again. The failure of the two Mackinnon Bills not only dashed the rather false optimism created by the 1835 Act, but indicated in a very obvious way that the law officers were not prepared to support radical alteration in the law. In 1839 Brougham's Act was amended for a minor technical reason,[1] and in 1844 the Judicial Committee of the Privy Council was empowered to extend patents up to a period of fourteen years.[2] Neither of these changes appears to have resulted from pressure applied by the invention interest. For most of the 1840s patent reform attracted little attention. Free trade and the revival of working-class unrest in 1842 and 1848 were the dominant domestic issues. Patent reform, by comparison, was unimportant, esoteric and dull. It was a subject for the hard-headed enthusiast, and demanded unfaltering attention rather than sparkling rhetoric. After the repeal of the corn laws and with the onset of an extreme depression, patent reform re-emerged as a major issue for inventors and manufacturers. Between 1848 and 1852 petitions from all parts of the country were presented to Parliament and the Board of Trade, and Patent Reform Committees, some organised on a national basis, sprang up in all the major industrial towns. In the late 1840s and early 1850s, when a significant proportion of an increasing number of patentees were still alive, reformers were determined to alter the law. In 1852 they managed, with the help of Brougham and Thomas Webster, to satisfy most of the demands which had been rejected for almost sixty years.

The setting up in 1847 of the National Association for the Reform of the Patent Laws marks the revival of the invention interest; Robert Scott Burn and John Hawden, two solicitors, presented the Association's petition to the House of Lords.[3] In the following year Henry

Archer, a gentleman from Pimlico, urged the appointment of another select committee,[4] and further petitions were presented by Henry Scrope Shrapnel, Thomas Grant, Thomas Prior, Frederick Campin, James Lea, Thomas Motley and William Whytehead, editor of the *Artisan Journal*.[5] The Society of Arts began discussions with Lefevre,[6] Secretary of the Board of Trade, and attempted to recruit the support of Charles Dickens and J. S. Mill.[7]

In Manchester the Patent Reform Committee which was first organised in 1828 now reformed itself under William Fairbairn and other manufacturers and engineers,[8] and in 1850 and 1851 a further eight reform associations sprang up: the Committee of the Society of Arts for Legislative Recognition of the Rights of Inventors; the Art Protection Society for the Amendment of the Laws affecting Letters Patent; the Birmingham Patent Law Reform Association; the National Patent Law Reform Association; the United Inventors' Association for the Amendment of the Law affecting Invention; the Patent Law League; the Association of Patentees and Proprietors of Patents for the Protection and Regulation of Patent Property; the Society for Promoting Scientific Inquiries with Social Questions: Reform of Patent Laws.[9]

This mounting pressure was not prompted solely by the conditions of trade, which in 1849 and 1850 showed signs of recovery, but by the excitement generated by the prospect of the Great Exhibition.[10] This provoked renewed interest in the patent question because manufacturers genuinely feared that unprotected exhibits would allow foreign competitors free access to the latest designs and technology. In December 1850 the Manchester Patent Law Reform Association took the opportunity to present a petition to H. Labouchère, President of the Board of Trade, who claimed that the subject was under consideration and that a Bill was being prepared. When no Bill appeared the Association appealed to Lord Brougham,[11] and he introduced an uncharacteristically ambitious measure, primarily for the purpose of 'forcing the government on to do something'.[12] His Bill was presented to the Lords on 24 March, and on 10 April Lord Granville, who was involved with the arrangements for the exhibition,[13] brought in the government's own draft.[14] Both Bills were referred to a Select Committee and the Board of Trade was ordered to prepare copies of the memorials and petitions it had received on the subject.[15] A separate Select Committee on Design was also established in the same year.

At first there were serious doubts about the intentions of the committee itself. The appointment of Granville as chairman — a well known opponent of the patent system — was seen by some as a devious move by the government and a prescription for doing nothing.[16] Even Granville was 'afraid that it might be thought he was taking a strange course when he supported the present bill, and at the same time avowed himself to be against the system'.[17] The *Mechanics Magazine* was incensed, and alluded to the 'private and sinister influences under which [the government Bill] had sprung so suddenly into existence'.[18] Several others were equally concerned, and Thomas Webster later noted that 'there are so many persons to be consulted, and so many interested in maintaining the system', that it was difficult to get anything done.[19] Granville's assurance that public opinion was not ripe for sweeping the system away probably allayed some fears; but everyone recognised that reform was not going to be without its difficulties. Even in 1851 reform was not a foregone conclusion.

The prospect of some reform brought in a flood of petitions from all parts of the country. Public meetings were arranged in Manchester, Birmingham, London and in almost every major industrial town. Societies, leagues, associations and reform clubs were organised with eagerness and anticipation. Patent reform was beginning to show signs of becoming a national movement of some significance. Everyone, it seemed, wanted to have a say. The members of the Wortley Mutual Improvement Society, of the Armley Youth's Guardian Society, of the Beeston Mutual Improvement Society and of the Churwell Mechanics' Institute,[20] were every bit as anxious to make their views known as the patent committees of Blackburn, Huddersfield, Bradford, Bridport,[21] Halifax,[22] Rochdale and Kirkcaldy,[23] and the chambers of commerce of Liverpool, Belfast and Manchester.[24] Equally anxious were the Inventors' Society, the Inventors' Aid Association, the London Mechanics' Institute,[25] the National Patent Law Amendment Association, the Inventors' Law Reform League, the Patent Reform Club and many others.[26]

The views and proposals expressed in these petitions were essentially the same as those expressed by reformers throughout the previous thirty years. There remained differences over detail but at least there was now a greater consensus: inventors deserved better protection; patents were too expensive; the procedure for granting them was archaic and cumbersome; there was a need, rather more tentatively expressed, for a Board of Examiners to reduce the number of trivial inventions. There were those who still disagreed with these central issues. The Association of Patentees for the Protection and Regulation of Patent Property opposed any preliminary 'inquisition' of specifications, as did petitioners from Lancaster and Leek.[27] The sugar refiners of Glasgow, Liverpool and London remained wholeheartedly against any reduction in the cost of patenting,[28] as did the Manchester Chamber of Commerce. In March 1851 it formed a sub-committee to examine a report from the Society for the Amendment of Patent Law, and concluded that:

> If the cost be made cheap, every trifling improvement, in every process of manufacture, would be secured by a patent; in a few years no man would be able to make such improvements on his machinery or processes as his own experience may suggest without infringing upon some other person's patent: useless litigation would follow and the spirit of inventions in small matters would be rather checked than encouraged.[29]

But these were the views of the dissenting few: in 1851 the majority of petitioners were fairly clear as to what should be done. When the 1851 Select Committee came to examine the Brougham and Granville Bills all the arguments were considered in depth.

Compared with the two earlier committees, the 1851 inquiry was a much more thorough affair. A large number of witnesses were called and, without exception, all were well informed. Every view was examined, including those of the abolitionists. The invention interest was represented both by individuals and by those closely associated with one or more of the various reform societies. Richard Roberts, I. K. Brunel, Charles May and others explained the difficulties they had encountered as individual inventors. William Fairbairn spoke on behalf of the Manchester Patent Law Reform Association, J. P. Westhead for the Inventors' Association, Henry Cole for the Society of Arts, and Frederick Campin explained the views of the United Inventors' Association and the Inventors' Patent Reform League. Patent agents were represented by William Carpmael, William Newton and William Spence, and patent officials were represented by several law officers and by Moses Poole. Thomas Webster provided most of the information on difficult legal matters (as J. Farey had done in 1829 and 1835) and several experts provided information on the various patent systems in Europe.

The Brougham and the Granville Bills had much in common, and according to Webster both were modelled on the recommendations first circulated by the United Association of Inventors in the autumn of 1850.[30] The principal objects were to lessen the delay and expense of patenting; to vary the duration of patents; to reorganise Patent

## The Patent Law Amendment Act

Office procedures; to set up a body known as the Commissioners of Patents; to allow protection from the date of application rather than from the date of the Great Seal; to allow inventors to enter a provisional specification to be amended within the first six months of the grant; to grant one patent for protection in England, Scotland and Ireland (previously patents were taken out separately for each country); and to provide an accessible index of patents together with a register of proprietors which was to enumerate names of assignees, shareholders of patents, licensees, and the districts in which the licences were enforceable.[31]

The major difference between the Bills concerned the appointment of examiners. Under the Brougham scheme the Commissioners of Patents (who were to consist of the Lord Chancellor, the Master of the Rolls, the Attorney or Solicitor General and the Lord Advocate for Scotland and Ireland) had powers to appoint a Board of Scientific Examiners to test the specification for novelty. The government Bill suggested that this function ought to reside with the Law Officers, who would seek the advice of scientists whenever required. Both views were widely discussed by witnesses examined by the committee. William Carpmael — the leading patent agent — supported the government Bill, since it left things much as they were.[32] He had a personal 'horror of legislating for practice' and felt that a permanent body of examiners would entail some erosion of the royal prerogative.[33] Archibald Rosser believed the American Board of Examiners was a miserable failure and saw no reason for the present system to be changed. The courts, he alleged, were the appropriate forum for testing the ultimate validity of patents.[34] Bennet Woodcroft could not imagine any satisfactory criteria for selecting examiners,[35] an argument which quite a number of witnesses accepted. Others, like Benjamin Fothergill, compromised. 'We in Manchester,' he explained, 'want the appointment of practical scientific men along with Officers of the Crown ... to examine all applications for patents.'[36] Thomas Webster also saw the examiners as an aid to the jurisdiction of the Law Officers, but on a much more permanent basis.[37] This appears to have been the general view of the reformers. Examiners would have powers of inspection, but would not replace the Law Officers.

For many, some sort of Board of Examiners was a necessary condition if inventors were to benefit from any decrease in the cost of patenting. Without it, the advantage of cheaper patents could too easily be offset by the expense of increased litigation. These two arguments went hand in hand.[38] Here again, though, there was disconcertingly little agreement among inventors on detail. The proposals for what a patent should cost were legion.[39] Most, though, were in favour of a system where fees would be paid in instalments and where the total cost for all three kingdoms would approximate to the cost of an English patent, that is, £100. The plan of the Committee of the Society of Arts, for example, proposed that inventors should pay £5 at registration, £10 at the first instalment, £20 at the second, £50 at the third and £100 as a final instalment, giving a total of £185.[40] Memorialists from Blackburn preferred something less: £10 on registration, £5 at the end of the first and second years, £20 after the third and £50 at the end of the seventh:[41] a total cost of £90. These differences were also reflected in the two Bills, although both accepted the need for payment by instalments. Brougham suggested a total cost of £140: £30 on registration, £40 after the third year and £70 after the seventh. Granville's Bill suggested a cost of £175, with three instalments of £20, £50 and £100 in the first, third and seventh years.

There was very little disagreement about paying fees in instalments. It allowed inventors to spread the cost of patenting, to relate patenting costs to the value of the invention itself and, finally, it 'put the thinking man in a better position to deal with capitalists' if he wanted to dispose of his invention.[42] On the actual cost of patenting there was wide disagreement, and old arguments were rehearsed repeatedly: costly patents prevented inventors from protecting their ideas, and cheap patents led to too many trivial inventions and increased litigation. Significantly, few inventors were against cheap patents. Most opponents were either Patent Office officials, patent agents or abolitionists.[43] William Carpmael, who was perhaps the most outspoken, even suggested that the matter of patent reform was something of a non-issue.

Although there was less disagreement on other issues, the difficulties of considering two very detailed Bills were becoming increasingly obvious. To simplify matters the committee eventually recommended the introduction of a third Bill, 'embodying as far as possible' the objects of the other two.[44] This Bill, which was introduced in June, was prepared by Thomas Webster. Not all the difficulties were removed, however. The *Mechanics Magazine* complained that it 'reflected too much the system he practised',[45] whilst the Society of Arts claimed that it 'virtually retained much of the cumbersome machinery'.[46] Moreover, the clause which suggested that prior publication in the United States and Europe was equivalent to public use at home provoked renewed opposition from the Manchester Patent

## The Patent Law Amendment Act

Law Reform Association[47] and from the Inventors' Aid Association, who at the same time warned that persistent opposition to the Bill might lead to yet another Select Committee, 'which may retard ... progress ... and throw Patent Law Amendment over to another session. Numerous inventors are now awaiting for a cheap patent which they have been kept in expectation of for two years.'[48] Similar fears were expressed by the Inventors' Society[49] and the Manchester Association, who urged acceptance of the Bill, despite its many faults, before the expiration of the Provisional Protection of Invention Act on 1 May 1852. They were prepared to accept what improvements there were 'rather than risk loss'.[50] Further petitions supporting the Bill now came from the London Mechanics' Institute, Blackburn, Bradford and Halifax, and from engineers and machinists and others from Rochdale. The National Patent Law Amendment Association also organised public meetings in London and Manchester to rally additional support.[51] In the event the Commons struck out the offending clause and rushed the Bill through at the last minute, pledging that the subject would be reconsidered in the near future.[52] On 1 July 1852 the Patent Law Amendment Act became law.[53]

The Act itself met most of the reformers' wishes. The administration of the system was simplified by setting up two offices, the Record and Commissioners' Office; inventors were protected from the time of application; a single patent for the United Kingdom replaced the system of separate patents for England, Scotland and Ireland; and an index of patents was to be drawn up and made accessible for public consultation. Finally, and most important, the cost of patenting was reduced to £180. Ironically, this was in excess of an English patent, so that most inventors would now pay more than they did under the old system. On the other hand, they now had protection in Scotland and Ireland and were able to spread the cost by paying £180 in three instalments over the first seven years of the patent's life. Significantly, the Law Officers retained the right to examine the technical merits of patent applications, but since few were ever assessed the patent system effectively remained a system of registration.[54]

The Act, which came into effect on 1 October 1852, was not greeted with jubilation, although most seemed happy enough with the progress that had been made. William Carpmael, who had been against any radical alteration, thought the Act a 'very substantial improvement'.[55] In a paper given to the British Association William Fairbairn praised the government for fully recognising the link between intellectual creativity and the 'products of national industry'.[56] Others were also impressed. The Society of Arts believed that the measure 'seems at last to offer sanguine hopes that the great leading principles so long advocated ... will be speedily realized in actual practice'.[57]

The *Mechanics Magazine* even saw fit to 'congratulate the invention interest on the position it [had] attained'.[58] And since almost all the reforming committees disbanded shortly after 1852 it seems that the general mass of inventors were reasonably satisfied.[59] There is little doubt that 'disappointment would have been widespread ... had the measure not been passed'.[60] The Act, though, was never regarded as a final solution, especially since the government had given a pledge to examine the subject again. Had the measure been seen as anything other than the 'foundation of the superstructure yet to be raised'[61] support would not have been so widespread.

The reform of the patent system was a slow process. There are several reasons why this was so. Firstly, the subject was inherently difficult. Patent laws were technically complex and intrinsically uninteresting. Many inventors were probably 'too ignorant to offer any interference',[62] and few MPs were willing or able to master the subject. In the 1820s Hume and Place declined their assistance for this reason, and when John Bright was approached by the Manchester Patent Law Reform Association in 1850 he assured them that a petition from men of science would have more effect than one from a politician.[63] Patent reform required 'zealous labour' in a 'toilsome and un-brilliant field'.[64] Secondly, the invention interest was not sufficiently unified, and remained organised on a local basis only, right through to the late 1840s. For the editor of the *Patent Journal* this was exasperating:

> until something like harmony activity and zealous exertion, are manifested by inventors, it is vain to expect any alteration in the patent law ... Inventors must unite ... Can no League be formed for the abolition of the Patent Laws? ... or must inventors, like the Jews, wait patiently for the coming of a most-delayed Messiah.[65]

Most other reform movements experienced similar difficulties in the first half of the nineteenth century,[66] but for the 'invention interest' this led to a proliferation of proposals which differed considerably in detail and sometimes in purpose. Harmony was not always evident, despite the existence of a consensus on broad themes. Finally, there was a 'want of sympathy on the part of the public with the comparatively small class of scientific and ingenious men [and] a mistaken jealousy' on the part of some capitalists who believed that reform would enhance the position of inventors at the expense of society. Taken together, as Webster observed, these 'gave power and effect to the obstructiveness or opposition of numerous persons in the three countries directly or indirectly interested in the official fees levied under the existing system'.[67]

Brougham's involvement in patent reform raises a final and more general issue: the link between Benthamism and the so-called mid-Victorian revolution in government. It has been argued by Fay, for example, 'that the majority of essential reforms accomplished between 1820 and 1875 had the Benthamite impress upon them'.[68] Although recent research, notably that of MacDonagh, Roberts and West,[69] has emphasised more institutional explanations, the influence of Benthamism has not been dismissed or ignored.[70] Brougham was in fact very much indebted to Bentham for his 'legal information' and once described the age of law reform and the age of Bentham as one and the same thing.[71] Significantly, Bentham considered the 'enormous expense of patents' as one of the three great burdens on 'men of industry',[72] but whether this influenced Brougham directly cannot be known, especially since inventors had been arguing the same from the 1820s. It is also difficult to assess whether the Acts of 1835, 1839, 1844 and 1852 had the Benthamite impress. Brougham was responsible for the first three and played a crucial role in passing the amendment Bill, but he generally relied upon the various proposals put forward by individual inventors or the invention interest. The 1851 Select Committee has been seen as a good example of Victorians attempting to legislate in the 'public interest on a high moral plane',[73] but whether it was part of some grand, perhaps utilitarian, scheme seems doubtful. Nor, on the other hand, is there any evidence to suggest that reform resulted from the spontaneous recognition of injustice (that is, intolerable evil) by those in authority. Law Officers and MPs were always cautious and generally accepted change rather grudgingly. In the end, patent reform depended upon the persistent and self-interested complaints of inventors, who were increasingly dissatisfied with the patent system and the protection which it provided. The 1852 Reform Act was their victory.

**Notes**

1  2 and 3 Vict., C.67; *Parl. Papers*, IV, 1839, pp. 371–3. This amendment merely prevented the Judicial Committee of the Privy Council from granting extensions for patents which had expired whilst the petition for prolongation was being considered.

2 7 and 8 Vict., C.69. This act appears to have resulted from an appeal by Lord Dundonald, the inventor. *See Hansard*, LXXII, 1844, pp. 1098–1103; *Mechanics Magazine*, XLI, 1844, p. 351; Brougham MSS, [18,012], Dundonald to Brougham, 4.7.1844; [24,223] 26.7.1844; [18,014] 14.9.1844; [18,015] 10.11.1844, U.C.L.
3 *L.J.*, LXXIX, 1847, p. 84.
4 *J.H.C.*, CII, 1848, p. 305.
5 *Public Petitions*, Appendix, 18th Report, Ref., 18,254 (App. 1,514) pp. 812–13; Ref., 10,226 (App. 1,091) pp. 496–7; Ref., 4,663 (App. 658) p. 350 (hereafter *Public Petitions*), H.L.R.O.
6 Society of Arts MSS, G. Grove to A. Lawrence, 30.11.1850. Ref., G.C.P. 51/29,80; Grove to Cabinet Ministers, 7.12.1850. Ref., G.C.P. 51/2a, 100; A. Aiken to G. Grove, 4.12.1850. Ref., G.C.P. 51/2a, 100; Wood, *op. cit.*, pp. 382–3; D. Hudson and K. W. Luckhurst, *The Royal Society of Arts, 1754–1954*, 1954. pp. 223–4.
7 *Select Committee on School Design*, Parl. Papers, XVIII, 1849, p. 32; *Fifty Years of Public Work of Sir Henry Cole K.C.B.*, 1884, II. pp. 273–5.
8 *Manchester Guardian*, 18.12.1850.
9 *Select Committee on Patents*, Parl. Papers, XXIX, 1864, p. 497.
10 *Select Committee on Patents*, Parl. Papers, X, 1871, p. 656; *Select Committee on Patents*, Parl. Papers, XVIII, 1851, p. 495; *Mechanics Magazine*, LIV, 1851, pp. 29–31.
11 *Manchester Guardian*, 31.1.1852.
12 *Select Committee on Patents*, Parl. Papers, X, 1871, p. 657; *Hansard*, CXV, 1851, pp. 2–3.
13 Lord E. Fitzmaurice, *The Life of Granville, George Leveson Gower, 2nd Earl Granville*, 1905, I, p. 41.
14 *L.J.*, LXXXIII, 1851, pp. 78, 125.
15 P.R.O. BT/1/483 737/51.
16 *Newtons Journal*, XXXI, Conjoined Series, 1851, p. 45.
17 *Hansard*, CXVIII, 1851, p. 13.
18 *Mechanics Magazine*, LV, 1851, p. 52.
19 *Select Committee on Patents*, Parl. Papers, XVIII, 1851, p. 204.
20 *Public Petition*, 1851, Ref., 11,836, (App. 1,506), p. 735; H.L.R.O.
21 *L.J.*, LXXIV, 1851, p. 20; *Public Petition*, 1851, Ref., 92, (App. 48) p. 25, H.L.R.O.; *Sessional Papers House of Lords*, XVI, 1851, pp. 458–9.
22 *L.J.*, LXXIV, 1851, pp. 24, 65, 82, 89; *Public Petition*, 1850, Ref., 1,916, (App. 175), p. 81; Ref., 4,245 (App. 383); Ref., 5,965 (App. 504) p. 226, H.L.R.O.
23 *L.J.*, LXXIV, 1851, p. 64; *Public Petition*, 1852, Ref., 5,494 (App. 836) pp. 420–1; *Public Petition*, 1851, Ref., 5,462 (App. 472), pp. 210–11, H.L.R.O.
24 *Public Petition*, 1851, Ref., 4,671 (App. 501), p. 234, H.L.R.O. Proceedings of the Manchester Chamber of Commerce, M8/2/5 p. 182, M.R.L.
25 *Public Petition*, 1852, Ref., 4,671 (App. 701), pp. 348–9; Ref., 4,922 (App. 737) pp. 366–7; Ref., 5,026 (App. 750), p. 373, H.L.R.O.
26 *Select Committee on Patents*, Parl. Papers X, 1871, pp. 200–1; *Sessional*

*Papers House of Lords*, XVI, 1851, pp. 455–8; P.R.O. BT/1/481/ 4,128–30;*L.J.*, LXXIV, 1851, p. 28; *Public Petition*, 1851, Ref., 11,444, (App. 1,418), p. 692; Ref., 4,522 (App. 426), p. 193; Ref., 93 (App. 49), pp. 25–6; Ref., 11,445 (App. 1,419), pp. 692–3; Ref., 11,446 (App. 1,420) p. 693; Ref., 7,471 (App. 709), p. 319, H.L.R.O.
27 *Public Petition*, 1851, Ref., 11,516, 11,904, 11,905, H.L.R.O.
28 *Public Petition*, 1851, Ref., 10,327 (App. 1,222), pp. 578–90, H.L.R.O.
29 Proceedings of the Manchester Chamber of Commerce, 1848–58, M8/2/5 p. 182, M.R.L.
30 Webster, *History...*, 1852, p. 10. R. H. Wyatt, editor of the *Repertory of Arts*, was the Honorary Secretary. He was examined by both the 1829 and 1851 Select Committees.
31 *Parl. Papers*, V, 1851, pp. 63, 91.
32 In his evidence before the 1833 committee John Farey claimed that he knew of only one or two cases where the Attorney General had refused to grant a patent for lack of novelty. The one Lord Denman refused in 1829 was the first instance he had known. He did note that several had been rejected since then. *Select Committee on Patents Amendment Bill*, 1835. In 1849 William Newton estimated that 3% of patent applications made in 1845 were refused. *Newtons Journal*, XXXIV, Conjoined Series, 1849, p. 47.
33 *Select Committee on Patents*, Parl. Papers, XVIII, 1851, pp. 239, 310.
34 *Ibid.*, pp. 549, 533.
35 *Ibid.*, p. 474.
36 *Ibid.*, p. 440.
37 *Ibid.*, pp. 57, 320.
38 See, for example, Webster's argument, *ibid.*, p. 57, and *Select Committee on the Design Act Extension*, Parl. Papers XVIII, 1851, p. 171.
39 *Patent Journal*, I–II, 1846, p. 246; *Public Petitions*, 1851, Ref., 6,294, (App. 521), p. 234, H.L.R.O.; Brougham MSS, [23,061], Reinagle to Brougham, 25.2.1837; [1,297] Dewey to Brougham, 23.3.1852, U.C.L.
40 *Mechanics Magazine*, LIV, 1851, pp. 6–7; *Select Committee on Patents* Parl. Papers, XVIII, 1851.
41 *House of Lords Sessional Papers*, XVI, 1851, pp. 458–9.
42 Society of Arts MSS, A. Aiken to G. Grove, 12.4.1850, Ref., G.C.P. 51/21/10.
43 *Select Committee on Patents*, Parl. Papers, XVIII, 1851, pp. 137, 155, 400, 403, 423.
44 *L.J.*, LXXIII, 1851, p. 274.
45 *Mechanics Magazine*, LIV, 1851, pp. 29–31.
46 *Public Petition*, 1851, Ref., 11,445 (App. 1,419), pp. 692–3, H.L.R.O.
47 *Manchester Guardian*, 31.1.1852.
48 *Public Petition*, 1852, Ref., 4,922 (App. 737), pp. 366–7, H.L.R.O.
49 *Public Petition*, 1852, Ref., 4,671 (App. 701), pp. 348–9, H.L.R.O.
50 *Manchester Guardian*, 31.1.1852.
51 *Public Petition*, 1852, Ref., 5,026 (App. 750), p. 373; Ref., 5,494 (App. 836), pp. 420–1; *L.J.*, LXXIV, 1852, pp. 20, 28, 66.
52 *Journal of the Society of Arts*, I, 1852–53, pp. 521–2; P.R.O. Granville Papers, 30/29/23/15/63. T. Lennard to Granville, 4.12.1852.

53 15 and 16 Vict., cap. 83.
54 The Patent Office was not given the power to examine patent applications until 1883 and then the Comptroller of Patents was only permitted to see whether the specification described the invention accurately. There was no test for novelty. This remained the case until 1907, after the Fry Committee had investigated the patent system. Novelty was then determined by searches through U.K. specifications for the previous fifty years.
55 *Select Committee on Patents*, Parl. Papers, X, 1872, p. 35.
56 *British Association for the Advancement of Science*, 28th Report, 1858, p. 164.
57 *Journal of the Society of Arts*, I, 1852–53, p. 522.
58 *Mechanics Magazine*, LV, 1851, p. 107. This judgement was made on the Webster Bill but obviously reflected what the *Magazine* thought of the Act generally.
59 The Manchester Patent Law Reform Association was still in existence in 1862. *Select Committee on Patents*, Parl. Papers, XXIX, 1864, pp. 497, 559.
60 *Hansard*, CXVIII, 1851, p. 185.
61 *Journal of the Society of Arts, op. cit.*, pp. 317–19.
62 Brougham MSS, [17,272], E. Watt to Brougham, 13.6.1851, U.C.L.
63 *Manchester Guardian*, 18.12.1850; Proceedings of the Manchester Chamber of Commerce, M8/2/5, pp. 217–18, M.R.L.
64 *Economist*, 19.7.1851.
65 *Patent Journal*, III, 1847, pp. 58–9.
66 Interest groups, with the exception of the railway interest, did not become organised on a national basis until the late nineteenth century. See G. Alderman, *The Railway Interest*, Leicester, 1973.
67 Webster, *History* ..., 1852, p. 5.
68 C. R. Fay, *Great Britain from Adam Smith to the Present Day*, 1928, p. 367.
69 O. MacDonagh, 'The nineteenth-century revolution in government: a reappraisal', *Historical Journal*, 1958, pp. 52–67; D. Roberts 'Jeremy Bentham and the Victorian administrative State', *Victorian Studies*, 1959, pp. 1–18; E. G. West, *Education and the Industrial Revolution*, 1976.
70 H. Parris, 'The nineteenth-century revolution in government: a reappraisal reappraised', *Historical Journal*, 1960, pp. 17–37; J. Hart, 'Nineteenth century social reform: a Tory interpretation of history', *Past and Present*, 1965, pp. 39–61.
71 S. E. Finer, 'The transmission of Benthamite ideas, 1820–1850', in G. Sutherland, ed., *Studies in the Growth of Nineteenth Century Government*, 1972, p. 11.
72 Stark, *op. cit.*, III, p. 524.
73 Harding, *op. cit.*, p. 9; Holdsworth, *op. cit.*, XIII, pp. 132–4.

# 4

# Patent law and the courts

In English patent law, inventors were given two closely related but quite distinct rights. Firstly, a patent gave inventors the right to *use* their invention as they pleased. Secondly, and more importantly, it gave them the right to *exclude* others from using the invention for a period of fourteen years, or longer if the patent was granted an extension. Neither of these rights was determined by the Patent Office. In the end, the value of patent property rights depended upon the decision made by the courts. 'Patents,' as Hindmarch explained, were 'a creature of the law, and the proprietor must depend almost entirely upon the law for vindication and support.'[1] The way in which judges interpreted patent law is, therefore, crucially important for assessing the value of this particular form of property right. In this chapter it will be shown that the quality of patent rights changed and was seen to have changed during the period 1750–1850. The main argument here is that from the mid-1830s the law, as applied by the courts, became both more certain and more favourable for patentees. The various problems which this caused inventors will be considered later.

Few would quibble with the notion that the certainty of the law is important for economic development.[2] Legal certainty ensures that contractual agreements enjoy a reasonably stable value over time, that the bundle of rights associated with private property can be used and transferred without increases in risk, and that interested parties know the legal consequences of their own actions in terms of costs and rewards. As industrial society grows in complexity and becomes increasingly interdependent, the need for a clearly defined and certain law becomes all the more important.

During the industrial revolution, however, the development of the law was strongly challenged and perhaps never more so than in the case of patent law. In the late eighteenth century the dismissal of Arkwright's patent created a great deal of confusion among inventors as to what the law really was.[3] In the evidence presented to the 1829 Select Committee this confusion was still evident, and almost all the witnesses condemned the lawmakers for undermining the value of patents. John Taylor thought that the uncertain nature of the law was the most objectionable aspect of the patent system. Joseph Merry, a ribbon manufacturer, claimed that he would sooner relinquish his patent than use the courts to settle the invasion of his rights,[4] whilst Samuel Clegg denounced the legal uncertainty despite successfully defending his 1815 patent.[5] By the early 1830s these opinions had not materially changed. Charles Babbage saw the patent laws as creating 'factitious privileges of little value', where 'the most exalted officers of the State in the position of a legalised banditto ... stab the inventor through the folds of an Act of Parliament and rifle him in the presence of the Lord Chief Justice of England'.[6] R. Thompson, writing in the *Westminster Review*, claimed that the patent laws 'have gloried in the shame of withholding their protection',[7] and John Farey dismissed the 'law proceeding on patent rights as a disgrace to our system of jurisprudence which ought not to be suffered to continue ... Appeal to the law,' he concluded, 'has proved itself to be a system of delusion.'[8] Although many patentees appeared to subscribe to the view that any questions relating to their inventions could easily be resolved by simply assuming that they were right and everyone else wrong, these reactions to the way in which the law was applied should not be ignored. In the late eighteenth and early nineteenth centuries the development of judge-made law was unpredictable and confusing.

One important cause of the observed uncertainty was the slow development of case law. This was a factor which could be removed only by time itself and by the increasing willingness of patentees to use the courts. Without a sufficient number of cases upon which precedents could be built up no measure of certainty, or at least no measure that would satisfy patentees, could be ensured. Few cases meant few precedents, and few precedents generally meant uncertainty. Moreover, with only a handful of cases, failures tended to become better known than successes, further restraining the patentees' inclination to litigate.

Down to 1800 there were few patent cases and they were rarely reported. In his memorial for the 1785 Committee of Patentees, Abraham Weston observed that:

> it may with truth be said that the [Law] Books are silent on the subject [of patents] and furnish no clue to go by, in agitating the Question What

is the Law of Patents? In the reports since Lord Mansfield has sat on the bench, there are not even titles 'Patent' of 'Monopoly' in the indexes to any of the reports of cases adjudged in his time; Tho' it is very well known that a great number of cases have been tried before him.[9]

Ten years later, Eyre, C.J., was still able to complain that 'patent rights are nowhere that I can find, accurately discussed in our books'.[10] This remained the case until the first decade of the nineteenth century.

From the data available it appears that in the second half of the eighteenth century no more than twenty-two cases came before the superior courts of London. The Dollands case in 1758, the first major case, was followed by four in the 1770s, nine in the 1780s, and eight in the 1790s. After the French wars the number increased. Between 1800 and 1810 eleven cases were reported and five of these came in 1803. Three more are reported for 1810 to 1815. There were eighteen cases between 1815 and 1820 and twenty-nine in the 1820s. In the early 1830s there were thirteen, but after 1835 a significant increase can be observed. From 1835 to 1840 there were thirty-four cases and in the 1840s the number increased to 128, nearly thirteen per annum.

These statistics do not bear close analysis. Not all London cases were reported and, without an extensive search through the assize records, there is no way of knowing how many cases were tried in the provinces.[11] Reporting may also have varied over time. But whilst the statistics are likely to understate the number of cases, they do suggest a turning point in the late 1830s and early 1840s. This is shown in table 2.

Table 2   The number of patent law cases, 1770–1849

|           | 1<br>Patents<br>granted | 2<br>Cases | 3<br>Patents<br>disputed | 4<br>2 as % 1 | 5<br>3 as % 1 |
|-----------|---------|-------|----------|---------|---------|
| 1770–1799 | 1,418   | 21    | 16       | 1·5     | 1·1     |
| 1800–1829 | 3,510   | 61    | 50       | 1·7     | 1·4     |
| 1830–1839 | 2,453   | 47    | 38       | 1·8     | 1·6     |
| 1840–1849 | 4,581   | 128   | 104      | 2·8     | 2·3     |

Source. B. Woodcroft, *Patents for Invention: Reference Index, 1617–1853*, 1855, pp. 655–81.

Even though the actual number of patents contested remained small, it is clear that the proportion coming before the courts increased. In one sense this is to be expected. An increase in the number of patents is likely to give rise to an increase in the number contested because of the growing interrelatedness of technology. But in another sense it shows an increased willingness on the part of patentees to use the courts to settle disputes over their property rights, especially after 1835 when the law (as will be shown) became more certain. It is important to note, however, that this increased willingness to use the courts in the second quarter of the nineteenth century was probably constrained, since, with a more certain law, patentees were able to sustain their rights more easily through *threats* of legal action rather than by legal action itself. By the 1830s each party knew the ground rules and how the courts would allocate the awards (the qualitative as opposed to the quantitative aspect of the statistics). Prior to the 1830s this appears not to have been the case. Success at law was believed to be more a matter of chance than of calculation, and because patentees were rarely willing to sacrifice their own inventions the small number of cases tended to perpetuate the observed uncertainty.

One major cause of the confused state of the law was the absence of a clear and unambiguous definition of what kind of invention could be the subject of a patent. According to section six of the 1624 Statute of Monopolies,[12] a patent could be granted only for the 'sole working and making of new manufacture', but neither judges nor inventors were clear what this meant.[13] In a period when the quantity and quality of inventive activity were changing rapidly, the meaning of 'new manufacture' as applied by the courts was often variable and obscure, and sometimes manifestly out of touch with the needs of a changing industrial economy.

This is clearly seen in Hornblower and Maberly *v.* Boulton and Watt (1799). Here Kenyon, C.J., defined 'manufacture' as something made by the hands of man'.[14] This indicated that the proper subject of a patent was a 'substance' or a 'thing made', but that is as far as it went. A more expansive though no more precise definition was given by Abbot, C.J., in 1819. He stated that 'the word manufacture has generally been understood to denote either a thing made, which is useful for its own sake or vendible as such; as a medicine, a stove, telescope, or some previously known article or in some other useful purpose, as a stocking frame, or a steam engine for raising water from mines'.[15] Although Abbot, C.J., like many of the judges, remained a 'stranger to the exact sciences' and lacked a practical knowledge of machinery,[16] he seems to have implied that the word manufacture could be either a new product or a new process to obtain well known ends. This 'double signification', as one writer in the *Westminster Review* noted, was not always admitted by some judges.[17]

The most serious difficulty judges encountered in defining manufacture arose when they struggled to distinguish between a manufacturing process, a method and a principle. According to the Statute of Monopolies a patent could not be granted for an abstract or philosophical principle; this amounted to patenting knowledge. Only when the principle could be carried out or reduced into practice was an invention entitled to the protection of a patent. The invention had to be corporeal, tangible and vendible.[18] Patents could not be granted — as Joseph Bramah was to say of Watt's invention[19] — for something that might be invented in the future. In the eighteenth century these conditions were rarely clear and some patentees defined their inventions in terms of a new principle simply to reduce the chance of infringement. Watt, for example, was advised by Dr Small to 'specify in the clearest manner that you have discovered some principles ... to secure you as effectively against piracy as the nature of the invention will allow'.[20] When Eyre, C.J., suggested in the 1776 Hartley case that a principle could be the subject of a patent these practices would appear to have been endorsed. But by the 1790s the emphasis had shifted. In the Watt case Abbot, C.J., held that a patent 'must be for some new production from these elements [principles] not for the elements themselves'.[21] He feared that patenting a principle would give inventors the sole rights to all future improvements. Collier, an early writer on patent law, argued much the same when he noted that a patent must be for 'some substantial thing produced'.[22] By the time of the 1829 Select Committee the matter had not been entirely settled. Although there existed a strong presumption that principles could not be patented, it was not until the 1830s and 1840s that a principle 'reduced into practice' was generally accepted as a proper subject for a patent.[23]

With regard to the question of 'method' judges were even less reassuring, and since 'two thirds or three quarters of the registered patents were for a method of operating, producing no new substance and employing no new machinery',[24] patentees were justifiably concerned. In the 1776 Hartley case Eyre, C.J., thought that a method was a proper subject for a patent, but when the matter was raised in

Watt's case against Bull, in the Court of Common Pleas, both Heath and Buller, J.J., expressed the opposite point of view, insisting that a new substance must be produced and that it must be vendible.[25] Not even Boulton's 'business dinner' with Buller, J., shortly before the delivery of the judgement shifted the balance of favour.[26] The issue of method was again raised when Watt brought a fresh trial in the King's Bench against 'your Hornblowers, your puff pastry ... [and] your Bristol Hogs'.[27] After their previous experience in the Court of Common Pleas Watt junior (who along with Boulton junior was given charge of this case) suggested that it would be more appropriate to define the invention as an addition and an improvement on the Newcomen and Smeaton engines.[28] He realised that there was little possibility of influencing Kenyon's, C.J., views by any 'out of doors arguments',[29] and knowing that the Judges of the King's Bench 'plainly intimated it [the method] would not be comprised under the term manufacture',[30] he thought there was only a slim chance of success unless his advised course of action was followed.

Watt's victory in 1799,[31] in which the 'doubts of the Courts of Common Pleas were treated with little ceremony',[32] did not mean, however, that the uncertainty concerning 'method' was resolved. Convincing the 'Gentlemen of the Long Robes that a steam engine is a tangible and vendible substance'[33] was far from convincing them that all other patents for methods or processes were of the same class and kind. Writing in 1806, William Hands stated that 'most of the patents now taken out, are by name, for the method of doing particular things: and where the patent is for only a method, if it be not affected or accompanied by a manufacture, it *seems* the patent is not good'.[34] In the 1817 case Hill *v.* Thompson, Eldon, L.C., suggested that a patent could be taken out 'even for a new method'.[35] But within two years Abbott, J., could go only so far as to say that the 'word manufacture *may, perhaps* extend to a new process to be carried on by known implements'.[36] In this case he treated processes and methods synonymously but, as the *Westminster Review* pointed out, this had not always been the case. When a method meant a new machine which could be sold it came within the definition of the 1624 Act; it was material and vendible. When a new method meant a new process which was not a vendible object some judges did not consider this as a 'new manufacture'.[37] By the 1830s the matter had become clearer. In 1834 James Russell managed to support his patent for a method (process), as did Derosne in the following year for his method of filtering the 'syrup of sugar'.[38] But according to William Hindmarch, the leading patent lawyer of the time, it was not until the 1842 case Crane v. Price that patentees could be said to have been fairly sure of what the courts meant,[39] namely that methods and processes were suitable subjects for a patent.

The area which undoubtedly caused most of the legal uncertainty was the specification. This was not required by the 1624 Act and, although James Nasmith's 1711 patent was the first to be described in this way,[40] it was not until 1734 that the Law Officers made it a provision in the patent itself.[41] Previously (as we have seen), patentees were only obliged to disclose to a limited number of tradesmen how the invention worked. There was no obligation to disclose generally information relating to the invention.[42] But in the eighteenth century, when the Law Officers found it necessary to distinguish between inventions and when it was felt necessary to encourage the diffusion of new skills among a wider community,[43] the specification became the accepted form of disclosure. This change from a contract between the patentee and the Crown to a 'social contract' between patentee and society was not given a secure legal footing until the 1778 Liardet v. Johnson case. Here Lord Mansfield concluded that 'the doctrine of the instruction of the public by means of the personal effects and supervision of the grantee was definitely and finally laid aside in favour of the novel theory that this function belongs to the patent specification'.[44] The doctrine was confirmed by Buller, J., in King v. Arkwright (1785), when he held that a patentee must:

> disclose his secret and specify his invention in such a way that others may be taught by it to do things for which the patent is granted, for the end and meaning of the specification is to teach the public after the term for which the patent is granted what the art is, and it must put the public in the possession of the secret in as ample and beneficial way as the patentee himself uses it. This I take to be clear law ... If the specification in any part of it be materially false or defective, the patent is against law and cannot be supported.[45]

In the late eighteenth and early nineteenth centuries a fair disclosure was deemed to have occurred so long as a man skilled in the particular trade could use the specification to make the subject of the patent. By the middle of the nineteenth century the emphasis had shifted to men of common understanding and with a 'moderate knowledge of the arts',[46] but never at any time were patentees obliged 'to teach any blockhead in the nation'.[47] Yet, whilst the law was clear, 'no branch of patent law has undergone' — so wrote Hindmarch — 'more discussion, or consideration than that relating to the specification and more patents have failed by reason of defects in their specification than from any other cause'.[48] There were a number of reasons why this should have been so.

In the late eighteenth century few inventors knew what a specification was meant to do.[49] This was certainly the case for patents taken out before the 1780s, but it becomes less true after the first Arkwright case. That this was so can be seen as early as 1781, when Matthew Boulton told Watt that 'your worthy ingenious and amiable friend R. Arkwright has totally lost his patent which he has renewed from time to time in new words and has entered a specification that no man can understand'.[50] 'Surely,' he later wrote, 'you cannot think it just that any tyrant should tyrannise over so large a manufactory by false pretences. He had no shadow of right and the judge and the whole court were unanimous that no mention was made in his specification of the thing or principle that was there in dispute.'[51] By the 1780s, then, it seems certain that most patentees had a good idea what the specification was meant to do,[52] although many failed to specify their invention adequately. This can be explained by the inability to describe fully what their invention was, and by their failure to distinguish between what was old and what was new. The growth of patent agents (as we shall see) in the second quarter of the nineteenth century considerably reduced the margin of error.[53] But since patents could be set aside because of any small ambiguity, or omission,[54] the drawing up of titles and specifications soon came to be regarded as a branch of metaphysics and one which patentees, understandably, found difficult to handle. Not all errors were unintentional, however, and there is little doubt that some patentees attempted to conceal as much as was possible to reduce the chance of imitation.[55] Some would also 'jumble' the specification with a number of different inventions to save taking out separate patents.[56] How many patentees deliberately drew up misleading specifications is difficult to assess, but since lawyers could easily discredit patents of this kind, it seems unlikely that inventors persisted in taking unnecessary risks.[57] It is undoubtedly true, however, that those specifications which attempted to deprive the public of information contributed, in part, to the uncertainty of which patentees complained.

Apart from the technical problems of interpreting the law, a great deal of the uncertainty seems to have arisen because of the excessively hostile attitude of some judges. According to one contemporary writer the 'courts of law looked with jealousy on the specification lest the bargain between the public and inventor should be too much in favour of the patentee'.[58] The fear that inventors would gain at the expense of the public good was, of course, a real one and was no doubt accentuated by those patentees who deliberately abused their rights. As a result judges tended to interpret the law in the strictest possible sense, allowing no error, however immaterial. Judges, as William Newton observed, scarcely even considered the patent as a whole. They usually seized upon some trivial and faulty point in order to show how patentees brought 'deceit upon the King'.[59] This often led to absurd consequences. Bainbridge's 1807 patent for *improvements* in flageolets was set aside because the one extra musical note only warranted a patent for an 'improved flageolet'.[60] In 1815 John Heathcoat withdrew his legal proceedings against Messrs Nunn, Brown and Freeman because the word 'bring' had been mistakenly inserted instead of 'put'.[61] William Newton also had to reapply for a patent on the same grounds; here the word 'pressing' had been used instead of 'dressing'.[62] These examples clearly indicate just how far the letter, rather than the spirit, of the law prevailed.[63]

Assessing the attitudes of judges towards patents is difficult. In the late eighteenth and early nineteenth centuries, when the law remained in an infant and unsettled state, judges rarely expressed a consistent viewpoint. They were seen to be at variance, not only with other judges but often with themselves.[64] Few were indiscreet enough to say what they may have felt. Lord Kenyon, C.J., was, however, fairly open in his views. Although totally unacquainted with mechanical subjects[65] he openly declard his opposition to patents. In Hornblower v. Boulton he stated, 'I'm not one of those who greatly favour patents,'[66] an opinion, according to one anonymous pamphleteer, 'which I have heard advanced for the sake of argument, but which I never expected to hear gravely delivered from the Bench'.[67] Unwilling to distinguish between patents and monopolies,[68] Kenyon found no difficulty in upsetting patent rights, and there were plenty who agreed with Francis Abbot when he claimed that 'in Lord Kenyon's time very little thing set aside a patent'.[69]

That Kenyon was not the only judge averse to patents is hinted at by a number of other judges. Buller, J., argued that 'many cases upon patents have arisen in our memory most of which have been decided against the patentee',[70] and Lord Eldon, who was himself reasonably fair with patentees, 'knew there were some sound opinions at variance with his own'.[71] Lord Ellenborough was another judge who did not agree with treating patentees hastily and thought inventors should be given the fruits of their labour.[72] By the early 1830s this view appears to have become much more general. In 1831 Lord Tenterden held that 'I cannot forbear saying that I think a great deal too much critical acumen has been applied to the construction of patents, as if the object was to defeat and not sustain them'.[73] Three years later Alderson, B., repeated these sentiments: 'We ought not to be too astute to deprive persons of the benefits to be derived from ingenious and new inventions.'[74] That Campbell — when he was Attorney General — should have seen Lord Tenterden as a 'strict judge with respect to patents'[75] suggests just how rapidly attitudes were altering during the 1830s. This change was summed up precisely by Baron Parke in the 1841 Neilson v. Harford case:

> half a century ago or even less, within the last fifteen or twenty years, there seems to have been very much a practice with both judges and juries to destroy the patent right by exercising great astuteness in taking objection, either as to the title of the patent but more particularly to the specification and many valuable patent rights have been destroyed in consequence of the objection so taken ... within the last ten years or more the courts have not been so strict in taking objection to the specification; and they have endeavoured to hold a fair hand between patentee and the public.[76]

The early prejudice against patents and the changing attitudes of judges are also supported by a statistical analysis of patent cases. From table 3 it is clear that judges were more generous to patentees after

Table 3 *Percentage of reported cases going for and against patentees at common law and at equity, 1750–1849*

|           | No. of cases | % of cases at common law | $FP_{CL}$ | $AP_{CL}$ | % of cases at Equity | $FP_E$ | $AP_E$ | % of unknown verdicts |
|-----------|------|------|------|------|------|------|------|------|
| 1750–1799 | 21   | 86   | 38·8 | 61·2 | —    | —    | —    | 14 |
| 1800–1829 | 61   | 69   | 30·9 | 69·1 | 18   | 45·5 | 54·4 | 13 |
| 1830–1839 | 47   | 53   | 76·0 | 24·0 | 21   | 60·0 | 40·0 | 26 |
| 1840–1849 | 128  | 49   | 76·2 | 23·8 | 28   | 55·5 | 44·5 | 23 |

$FP_{CL}$ For plaintiff at common law.
$AP_{CL}$ Against plaintiff at common law (including writs of *Scire facias*).
$FP_E$ For plaintiff at Equity.
$AP_E$ Against plaintiff at Equity.

*Source.* Woodcroft, *op. cit.*, 1855, pp. 655–81.

the 1830s than they were in any previous period. Between 1750 and 1799 only 39% of common law cases were in favour of the patentee; this had increased to 76% by the 1840s. Although in part it can be explained by patentees improving their specifications (see next chapter) to meet the rigorous requirements of the courts, there is equally very little doubt that the statistics reflect what judges felt about patents for invention.[77]

It seems that patentees, or at least their agents, were fairly astute in recognising the changing attitude of the courts. In his evidence before the 1829 Select Committee John Farey had been critical of the courts, but by 1835 he had modified his views.

> A great alteration has taken place since ... [the 1829 Select Committee] ... in the feeling of the Court of Law towards patent rights and hence some of the observation I made then, I should now modify because many of the grievances of which patentees then complained loudly, have not been felt so much since, in consequence of the courts exercising in a much more favourable manner towards patentees than they had done previously.[78]

Even the widely read *Mechanics Magazine*, normally so critical of the patent laws, conceded in 1836 that the

> decided turn which the feelings of *Judges, Jurors and the public* have taken in favour of inventors ... [suggests] ... a better state of things is evidently approaching. The verdict which so righteously secured Messrs. Mackintosh and Co.'s patent is *another* and decisive blow, both at the old monstrous legal theory of patents and at the prevalence of the robber-like maxim, which has too long influenced many in the manufacturing world.[79]

And by the late 1840s Barlow, the editor of the *Patent Journal*, was arguing that 'the rights of patentees are so rightly upheld, so favourably regarded by the judges of the land, that we know of no other description of property which is so secure'.[80] By the time of the Great Exhibition there were few who complained about the attitude of the courts. Most patentees thought the law clear and fairly balanced.[81]

Judges' attitude to patents changed because they now accepted that inventions led to prosperity and economic growth. The early prejudice against labour-saving machinery, which according to one writer 'found its firmest stronghold on the Bench',[82] had by the 1830s almost disappeared: patents were useful, important and necessary for the growth of industry. This view was not entirely a nineteenth-century phenomenon. David Hartley's 1773 patent for fire plates is a case in point. When the patent came before the courts in 1776,[83] Eyre, J., supported it on the grounds that 'the invention consisting in the method of disposing those plates of iron so as to produce their effect, and that effect being a *useful and meritorious one*, the patent seems to have been properly granted'.[84] Utility also played a part in Watt's case against Hornblower, when Ashurst, J., argued that

> the encouragement of new inventions is of infinite importance to the Kingdom; and where it appears incontestable that the invention does possess all the useful properties that this ... [steam engine] ... professes to do, it would be hard if the inventor should be robbed of his reward by a frivolous criticism, when the public are actually enjoying the fruits of his labour ...[85]

But if in the late eighteenth century judges were beginning to use consequential arguments as a means of supporting patents, it seems that the practice was not widespread.[86] In fact, in the 1781 Arkwright *v.* Nightingale case Lord Loughborough warned that 'we must *never* decide private rights upon any idea of public benefit; a cause between individuals cannot be determined upon consequential reasons'.[87] Even as late as 1836 Baron Parke doubted 'whether the question of utility is anything more than a compendious mode, introduced in comparatively modern times of deciding the patent to be void under the Statute of Monopolies'.[88] Yet, in spite of these doubts, the question of relating patented inventions to economic growth had by the 1830s secured legal footing. 'It is of great importance,' the Solicitor General quite typically noted,

> that due encouragement should be given to talent, and to genius and to industry, and, where it takes place, the expenditure of capital ... We are all interested that fair protection should be given to objects of this nature.[89]

By the 1830s, it was 'scarcely necessary ... to suggest that patents were amongst the most effective modes of advancing the best interests of society'.[90]

The emphasis that judges were putting on consequential effects was not something unique to patents,[91] nor was the idea of utility a new one. In Elizabethan times patents had been justified in exactly the same way. But in the nineteenth century, when the scriptures of political economy and the utilitarian creed were in the ascendancy,[92] and when economic growth forced itself on the contemporary mind, judges were unconsciously led to transpose theoretical constructs into practical reality. Few judges could resist extolling the 'economic benefits of machinery' and few missed the opportunity to lecture on the virtues of political economy.[93]

Writing in 1877, Robert Macfie suggested that the 'history of patents is one of continually relaxing aversion on the part of the Courts of Law'.[94] Although this was obviously a slow process, much of the change took place between 1830 and 1840. In this period the courts created a new equilibrium wherein the difficult problem of reconciling individual justice with the needs of public liberty and the needs of an expanding economy had, especially from the patentees' point of view, been largely resolved. Whether the judges were correct in assuming that patents were important for economic growth, and whether they were correct in reallocating rights between inventor and society, is another matter, which cannot be answered here (if it can be answered at all). All that need be noted is that a change in the attitude of judges did take place, that it occurred some time in the early 1830s, and that patentees believed the value of their property rights was now more certain and secure.

## Notes

1 W. Hindmarch, *Law and Practice of Letters Patent for Invention*, 1848, p. 248.
2 F. A. Hayek, *The Constitution of Liberty*, 1960, p. 208.
3 See Robinson, *op. cit.*, 1971, where this is examined in detail.
4 *Select Committee on Patents*, Parl. Papers, III, 1829, pp. 13, 89, 96.
5 *Crossley* v. *Beverly*, 1829, T. Webster, *Reports and Notes of Patent Cases*, 1844, I, pp. 106–20 (hereafter, *Reports*); W. Carpmael, *Report on Patent Cases*, 1849, I, pp. 480–8 (hereafter, *Reports*).
6 *Quarterly Review*, XLIII, 1830, p. 333.
7 *Westminster Review*, XXII, 1835, p. 471.
8 Brougham MSS, [45,493], Farey to Brougham, 21.8.1833; see also J. Prince to Brougham, [43,768, 45,198], 1.4.1833, 30.7.1833, U.C.L.
9 Boulton and Watt MSS, *Observations on Patents*, Parcel E, B.R.L. Weston is inclined to exaggerate the number of cases before Lord Mansfield. In *Liardet* v. *Johnson* (1778) Lord Mansfield claimed in his summing up that he had deliberated on patents in several but not many cases. See E. Wyndam Hulme, 'On the history of patent law in the seventeenth and eighteenth centuries', *Law Quarterly Review*, 1902, p. 295. Prior to the mid-eighteenth century patents came under the jurisdiction of the Privy Council, but in 1753 all questions concerning patents were left to the common law courts. A. Harding, *A Social History of English Law*, 1966, p. 313.
10 Boulton and Watt MSS, *The Special Case in the Cause of Boulton and Watt* v. *Bull C.C.P. 1795*, Parcel E, p. 19 B.R.L. Hindmarch states that up until the reign of George III 'our law reports ... are almost entirely silent respecting such patent privileges', *op. cit.*, p. 6.
11 Woodcroft provides the major sources for patent cases before 1852. It is a reasonably exhaustive series and is arranged alphabetically. None of

the cases is dated. How many cases escaped his survey is difficult to estimate, but his sample appears to represent a very large proportion of them. This is corroborated by the evidence gathered by John Farey for the 1829 Select Committee (Appendix B) and by an examination of a large number of contemporary law reports. These statistics include cases tried in Scotland. According to T. Webster there were probably only twenty cases between 1750 and 1852, *Select Committee on Patents*, Parl. Papers, XVIII, 1851, p. 122. Some patents came before the courts more than once.

12 21 James I, cap. 3.2.6.
13 R. Godson, *A Practical Treatise on the Law of Patents*, 1823, p. 57.
14 Carpmael's *Reports*, I, p. 158.
15 *King* v. *Wheeler*, 1819, Carpmael's *Reports* I, p. 397, *Select Committee on Patents*, Parl. Papers, III, 1829, p. 108.
16 J. L. Campbell, *The Lives of the Chief Justices of England*, 1874, IV, p. 319. See also *Select Committee on Patents*, Parl Papers, III, 1829, p. 45; Brougham MSS, [28,617], A. Galloway to Brougham, 'It is a great misfortune that there are but few persons at the Bar who have any extensive practical knowledge of machinery', 1.3.1825, U.C.L.
17 *Westminster Review* XXII, 1835, p. 457.
18 Godson, *op. cit.*, p. 83.
19 J. Bramah, *A Letter to the Right Hon. Sir James Eyre L.C.J. C.C.P. on the subject of the cause Boulton and Watt* v. *Hornblower and Maberley for infringement on Mr. Watt's Patent for an Improvement on the Steam Engine*, 1797.
20 Robinson, *op. cit.*, 1971, p. 120.
21 Godson, *op. cit.*, 1840 ed., p. 74.
22 J. D. Collier, *Law of Patents*, 1803, p. 79.
23 *Select Committee on Patents*, Parl. Papers III, 1829, p. 107; *Horsehill and Iron Co.* v. *Neilson and Others*, 1843, Webster's *Reports*, I, pp. 677–720.
24 Boulton and Watt MSS, *The Special Case in the Cause Boulton and Watt against Bull in Court of Common Pleas*, 1795, p. 23, Parcel E, B.R.L.
25 Robinson, *op. cit.*, 1971, p. 123.
26 J. P. Muirhead, *The Life of J. Watt*, 1911, p. 405. Muirhead saw the meeting as one where Buller 'was dressing with flowers the victim he was preparing to sacrifice'. It is also worth noting that Watt had also written to Justice Eyre prior to the trial. Boulton and Watt MSS, Watt to Eyre, March 1795, Parcel E, B.R.L.
27 Boulton and Watt MSS, Boulton to Watt junior, 28.11.1796, Parcel D, B.R.L.
28 Boulton and Watt MSS, Watt junior to Boulton, 4.12.1798, Parcel E, B.R.L.
29 Boulton and Watt MSS, Watt junior to Watt, 22.11.1798, Parcel É, B.R.L.
30 Boulton and Watt MSS, Watt junior to Watt, 4.12.1798, Parcel E, B.R.L.
31 Boulton and Watt MSS, Watt junior to Boulton junior, 25.1.1799 Parcel E, B.R.L.

## Patent law and the courts

32 Boulton and Watt MSS, Watt junior to W. Lawson, 26.1.1799, Parcel E, B.R.L.
33 Boulton and Watt MSS, Watt junior to Boulton junior, 8.2.1799, Parcel E, B.R.L.
34 W. Hands, *The Law and Practice of Patents for Invention*, 1806, p. 6.
35 *Hill v. Thompson*, 1817, Webster's *Reports*, I, p. 237.
36 *King v. Wheeler*, 1819, *op. cit.*
37 *Westminster Review*, XLII, 1835, p. 457.
38 Russell MSS, Bundle No. 16, 67/15/22. *Russell v. Cowley* (Legal Papers), Stafford County Record Office, William Salt Library; Webster's *Reports*, I, pp. 455–72; *Derosne v. Fairrie* (K.B.), 14.2.1835. Webster's *Reports*, I, pp. 154–66. John Fairrie claimed that, although the judge was in his favour, the jury were not. *Select Committee on Patents*, Parl. Papers, XVIII, 1851, p. 148.
39 Hindmarch, *op. cit.*, p. 84.
40 D. Seaborne Davies, 'Early history of English specifications', *Law Quarterly Review*, 1934, p. 272; Gomme, *op. cit.*, 1946, p. 25.
41 Searborne Davies, *op. cit.*, p. 90.
42 L. Getz, 'A history of the patentee's obligation in Great Britain', *Journal of the Patent Office Society*, 1964, pp. 69–73 (hereafter *J.P.O.S.*).
43 J. Prager, 'Early growth and influence of intellectual property', *J.P.O.S.*, 1964, p. 288; Gomme estimates that of the 158 patents registered between 1711 and 1734 only twenty-nine had specifications, *op. cit.*, p. 27; Macleod, *op. cit.*, pp. 40–1, 43–5.
44 Wyndham Hulme, *op. cit.*, p. 317. For a discussion of the evidence on this point see Robinson, *op. cit.*, p. 119 n. 13.
45 *The Trial of a Cause instituted by R. P. Arden Esq., His Majesty's Attorney General, by a Write of Scire Facias, to repeal a Patent granted on the 16th December 1775 to Mr. Richard Arkwright*, 1785, p. 172.
46 Godson, *op. cit.*, p. 118.
47 Small to Watt, 5.2.1769; cf. Musson and Robinson, *op. cit.*, 1969, pp. 54–6.
48 Hindmarch, *op. cit.*, p. 158.
49 Robinson, *op. cit.*, p. 1971.
50 Boulton and Watt MSS, Boulton to Watt, 24.7.1781, Parcel D, B.R.L.
51 Boulton and Watt MSS, Boulton to Watt, 7.8.1781, Parcel D, B.R.L.
52 It appears that Wedgwood, as early as 1775, may have been aware of what the specification was meant to do. In attacking Richard Campion's application for an extension Wedgwood argues that 'no new Art is taught to the *public*'; cf. Robinson, *op. cit.*, 1964, p. 212.
53 *Bovill v. Moore*, 1816; Carpmael's *Reports*, I, p. 339. Godson, *op. cit.*, p. 118.
54 *Select Committee on Patents*, Parl. Papers, III, 1829, p. 18.
55 *Barber v. Grace*, 1847; *Newton's Journal*, XXXI, Conjoined Series, 1847, p. 435.
56 *Morgan v. Seaward*, 1836; Webster's *Reports*, I, p. 173; J. Davies, *Patent Cases*, 1816, p. 27.
57 See A. Rosser's evidence given before *The Lords Select Committee on Patents Amendment Bill*, 1835, H.L.R.O.

58 Godson, *op. cit.*, pp. 107–8.
59 *Select Committee on Patents*, Parl. Papers, III, 1829, p. 176.
60 *Westminster Review*, XXII, 1835, p. 447.
61 W. Felkin, *History of the Machine-wrought Hosiery and Lace Manufactures*, Newton Abbot, 1967 ed., p. 210.
62 *Select Committee on Patents*, Parl. Papers, III, 1829, p. 70.
63 Lord Brougham's 1835 Act allowed errors of this kind to be changed by a disclaimer in a memorandum of alteration. See above.
64 *Select Committee on Patents*, Parl. Papers, III, 1819, p. 110.
65 Boulton and Watt MSS, Watt junior to Boulton junior, 25.1.1799, Parcel E, B.R.L.; See also *The Life of Lloyd, the first Lord Kenyon L.C.J. of England*, 1873; *A Sketch of the Life and Character of Lord Kenyon*, 1802.
66 Godson, *op. cit.*, p. 204.
67 *Observations, op. cit.*, 1791. This comment was prompted by one of Kenyon's earlier but unrecorded judgements.
68 *Hansard*, XV, 1833, p. 977.
69 *Select Committee on Patents*, Parl. Papers, III, 1829, p. 64.
70 *Turner* v. *Winter*, 1787, cf. Hindmarch, *op. cit.*, p. 16.
71 *Beaumont* v. *George*, 1815, Carpmael's *Reports*, I, p. 303.
72 Campbell, *op. cit.*, IV, p. 196; *Hansard*, XV, 1833, p. 977.
73 *Hullet* v. *Hague*, 1831, Carpmael's *Reports*, II, p. 356.
74 *Russell* v. *Cowley*, 1834, Hindmarch, *op. cit.*, p. 197.
75 *Cornish* v. *Keene*, 1837, Webster's *Reports*, I, pp. 501–12.
76 *Neilson* v. *Harford*, 1841, p. 138, B.M.
77 See Pollock in *Kay* v. *Marshall*, 1836, 'I am quite sure that I am correct in stating that the modern views of the learned judges ... [are] somewhat different from what was the case ... [and] I take it that the view is now taken of a patent is, that, in *whatever language it may have been described, however clumsily it may have been expressed*, if the patent is an improvement it will be sustained.' Webster's *Reports*, I, p. 138.
78 *The Lords Select Committee on Patents Amendment Bill*, 1835, H.L.R.O.
79 *Mechanics Magazine*, XXIV, 1836, p. 460.
80 *Patent Journal*, III, 1847, p. 271.
81 *Select Committee on the Design Act and Extension Bill*, Parl. Papers, XVIII, 1851, pp. 25, 37; *Select Committee on Patents*, Parl. Papers XVIII, 1851, pp. 191, 233, 279. *Mechanics Magazine*, LIV, 1851, p. 7; T. Turner, *Copyright in Design in Art and Manufactures*, 1848, p. 15; For some exceptions see *Public Petitions*, 1851, Ref. 11,444 (App. 1,418), p. 692; Ref. 4,522 (App. 426), p. 193.
82 J. Coryton, *A Treatise on the Law of Letters Patent*, 1855, p. 54.
83 Hartley MSS, 'An account of the invention and use of fire plates', 23.9.1785, pp. 1–16. D/EHY 88 17, Berkshire County Record Office, (hereafter B.C.R.O.).
84 Quoted in *Huddart* v. *Grimshaw*, 1803, Webster's *Reports*, I, p. 230. See also Hartley MSS, Rayley to Hartley, 12.2.1777; W. Bathoe to Savile, 30.11.1776. D/EHY Acc. No. 644 B4/2, B.C.R.O.
85 Boulton and Watt MSS, *Arguments of the Judges in Mr. Watt's Patent Trial*, 1799, p. 72, Ref. M.I., B.R.L.

86 Coryton, *op. cit.*, p. 38.
87 *Select Committee on Patents*, Parl. Papers, III, 1829.
88 *Morgan* v. *Seaward*, 1836, Webster's *Reports*, I, pp. 166, 170, 187.
89 *Crane* v. *Price and Others*, 1840, Webster's *Reports*, I, p. 377. See also *Regina* v. *Cutler*, 1847, *Patent Journal*, IV, 1847, p. 62.
90 *Hansard*, XXXVI, 1837, pp. 554–5.
91 It also appears to have been by the mid-nineteenth century an important factor in the development of nuisance law. See J. F. Brenner, 'Nuisance law and the industrial revolution', *Journal of Legal Studies*, 1974, pp. 403–33.
92 M. Blaug, *Ricardian Economics*, New Haven, Conn., 1958, chs. 3, 7.
93 J. L. and B. Hammond, *The Village Labourer*, 1966 ed., p. 275; R. K. Webb, *Harriet Martineau: a Radical Victorian*, 1960, p. 102.
94 Society of Arts, MSS, GCP 18/9C – G, 1870, p. 49.

# 5

# Patent agents: the early growth of a nineteenth-century service sector

In 1850 William Newton observed that 'patent property has become of enormous value within the last few years [and] it has called forth a class of men that occupy an intermediate position'.[1] Although Newton does not bother to say who these men were or what they did, he was referring to patent agents. By 1851 they were an established part of the scene and roughly 90% of all patents granted passed through their hands.[2] The purpose of this chapter is to analyse the growth of this service sector[3] and to see how it created an important link between inventors, manufacturers, the law and the administration of the patent system.

The emergence of the patent agent was a nineteenth-century phenomenon.[4] In the eighteenth century the number of patents and court cases did not warrant the growth of a specialised advisory service for inventors. In the 1780s and 1790s, when the law was seen to be uncertain, inventors obviously sought the best legal advice,[5] but despite Watt's exasperated claim that 'it is not to be supposed that every inventor is lawyer enough to know what sense the Court of Justice have put upon [a] clause in the specification',[6] few men specialised in this area. Most inventors generally relied upon the assistance of friends and solicitors to help with patent applications, specifications and other legal problems.[7] But as the annual number of patents and court cases increased, the changing structure of the invention industry created the opportunity for some to make a living as fee-earning consultants, men 'who purveyed skilled advice and direction rather than produced or sold goods'.[8]

The first patent agents were Patent Office officials. From the scant evidence, it appears that James Poole was the first to advise inventors on patent applications. As Clerk for Invention between 1776 and 1817 he was also to combine private and public business. There are no records to indicate the extent of his business, but his son, Moses Poole, reported in 1849 that his 'father always acted as a patent agent'.[9] Moses Poole, who took over his father's position in 1817, was, however, to become one of the most prominent patent agents in the unreformed period.[10] In 1821 he formed a partnership with the engineer William Carpmael, who had set up on his own account two years earlier,[11] and by 1829 their business was flourishing.[12] William Newton, who himself started up as a patent agent in 1819 when employed as a draftsman in the Enrolment and Rolls Chapel Office (both in the Patent Office), claimed that Poole had almost a monopoly of the patent agent business.[13]

The formation of the Poole–Carpmael partnership was regarded by later patent agents as the start of the profession,[14] and there is little doubt that they, and Newton, controlled most of the business in the 1820s and 1830s. Joseph Clinton Robertson, who founded the *Mechanics Magazine* in 1823, was the only newcomer in the 1820s.[15] In the 1830s, though, the number of patent agents slowly increased. One of Robertson's former associates at the *Magazine*, Fredrick William Campin, engaged in the patent business in 1834[16] and in 1839, seven years after having been articled to William Newton, William Spence branched out on his own account.[17] By 1840 Miles Berry, another Patent Office official, had joined William Newton as a partner, and Luke Herbert and Thomas Gill set up their own practices. Over the next ten years the number of patent agents appears to have doubled. In the 1851 *Post Office London Directory* the following newcomers are noted: Abbot and Wheatley, W. Baddeley, Barlow, Payne and Parker, J. Bethall, R. A. Brooman, C. Cowper, H. Dircks, C. Dod, J. Gedge, Green and Prince, Greville and Hard, and J. C. Haddan. Three agents were omitted from the list: J. Bates, B. Rotch and B. Woodcroft. W. E. Newton and A. V. Newton both worked for their father, W. Newton. According to Campin only about half of these acted as full-time patent agents, and some were simply concerned with registrations of designs. He also confirms that the old partnerships, like Carpmael, Poole Newton and Co., and Robertson, still dominated the business.[18] In fact in 1851 Carpmael claimed he had 'more than half of the current enquiries for patents'.[19]

There are two significant features of this growth in patent agents. Firstly, and predictably, all were located in London, near the Patent Office. Luke Herbert, who dealt largely with inventors in Birmingham, and Bennet Woodcroft, who originally practised in Manchester, are the only two who appear to have started in the provinces. Both eventually moved to London because there was insufficient invention business in either of the two industrial cities to provide an adequate income. Richard Prosser estimated that only 557 patents were taken out by Manchester inventors between 1723 and 1851, and in a similar estimate for Birmingham he records 755. Even in the prosperous 1840s the annual number of patents in either city did not exceed forty-five. Only in London was it possible to make a living.

The second significant feature is that most patent agents — apart from those who were Patent Office officials — were generally practising engineers, who came to know the law through practical experience. Evidence on individual agents is difficult to come by, and although many of the leading agents were examined by the 1851 Select Committee they said very little about their background. William Carpmael was certainly an engineer of some standing and intended to make his practice 'like a manufactory of patents'. In Crane *v.* Price (1842) he revealed that throughout his life he had advised on 'building structures and machinery of every class and kind' and that he was at one time the chief engineer to a Cheshire salt works.[20] When he acted in these various capacities is not known, but since he was elected a member of the Institute of Civil Engineers in 1828 it would appear that his engineering practice was well established before he formed his partnership with Moses Poole.[21]

Another engineer with a well established practice in the 1820s was John Farey. Although he was later to dissociate himself from the growing class of patent agents, Farey was described by Prosser as the '*facile princeps*' at drawing up specifications.[22] Born in 1791, he was part of the Scottish intellectual invasion of England during the industrial revolution. From the age of fourteen, when he was making illustrative plates for the *Rees* and *Edinburgh* encyclopaedias, up to his death in 1851, he is to be found trying his hand with mixed success at inventing, engineering, consulting, writing and lace manufacturing.[23] In fact, it was in lace manufacturing that Farey started up his patent business. In 1817 he prepared John Heathcote's specification,[24] after the one drawn up by R. Blunt and C. Staveley[25] proved impossible to defend in the courts. In 1819 Farey also acted on behalf of Mr Hall of Basford, another lace manufacturer, who had invented a process of singeing lace net by means of a gas flame. By 1826 Farey had become a full-time consultant engineer. In the following year he published his *Treatise on the Steam Engine* and in 1829 and 1835 gave evidence to the two Select Committees on patents. None of this

*Table 4   Number of patents taken out in Manchester and Birmingham, 1723–1852*

|  | M | B |  | M | B |  | M | B |
|---|---|---|---|---|---|---|---|---|
| 1722 |   | 1 | 1790 |   | 7 | 1821 | 2 | 6 |
| 1723 | 1 |   | 1791 |   | 1 | 1822 | 4 | 3 |
| 1738 |   | 1 | 1792 |   | 3 | 1823 | 8 | 7 |
| 1742 |   | 1 | 1793 | 2 |   | 1824 | 5 | 12 |
| 1748 | 1 | 1 | 1794 | 1 | 3 | 1825 | 7 | 12 |
| 1759 |   | 2 | 1795 |   | 3 | 1826 | 4 | 9 |
| 1762 | 1 | 1 | 1796 | 2 | 5 | 1827 | 4 | 11 |
| 1763 | 1 |   | 1797 |   | 1 | 1828 | 3 | 5 |
| 1765 | 1 |   | 1798 | 1 | 6 | 1829 | 3 | 9 |
| 1768 |   | 3 | 1799 | 2 | 2 | 1830 | 3 | 9 |
| 1769 |   | 2 | 1800 | 2 | 5 | 1831 | 5 | 10 |
| 1770 | 2 | 2 | 1801 | 2 | 3 | 1832 | 4 | 12 |
| 1771 |   | 1 | 1802 | 3 | 3 | 1833 | 9 | 12 |
| 1772 | 2 | 2 | 1803 | 3 | 1 | 1834 | 14 | 8 |
| 1773 |   | 1 | 1804 | 1 | 4 | 1835 | 14 | 16 |
| 1774 | 1 | 1 | 1805 |   |   | 1836 | 19 | 17 |
| 1775 | 1 | 1 | 1806 |   | 3 | 1837 | 22 | 16 |
| 1776 | 1 | 1 | 1807 |   | 2 | 1838 | 18 | 22 |
| 1777 |   | 1 | 1808 |   | 8 | 1839 | 31 | 24 |
| 1778 |   | 3 | 1809 | 3 | 6 | 1840 | 25 | 33 |
| 1779 |   | 4 | 1810 | 2 | 6 | 1841 | 20 | 34 |
| 1780 | 1 | 5 | 1811 |   | 11 | 1842 | 20 | 29 |
| 1781 | 1 | 4 | 1812 | 1 | 11 | 1843 | 23 | 22 |
| 1782 | 1 | 1 | 1813 | 1 | 10 | 1844 | 21 | 24 |
| 1783 | 2 | 2 | 1814 | 1 | 6 | 1845 | 25 | 29 |
| 1784 |   | 1 | 1815 | 6 | 2 | 1846 | 21 | 30 |
| 1785 |   | 3 | 1816 | 3 | 5 | 1847 | 27 | 22 |
| 1786 | 1 | 8 | 1817 | 1 | 6 | 1848 | 23 | 23 |
| 1787 | 1 | 6 | 1818 | 5 | 5 | 1849 | 30 | 30 |
| 1788 | 2 | 1 | 1819 | 1 | 8 | 1850 | 44 | 22 |
| 1789 |   | 1 | 1820 | 4 | 7 | 1851 | 31 | 30 |
|  |   |   |  |   |   | 1852 | 30 |   |

*M* Manchester, *B* Birmingham.

*Source.* H. G. Prosser, *An Appreciation of R. B. Prosser*, MS/608/p. 32, M.R.L.; M. Smith, 'Patents for invention: the national and local picture', *Business History*, 1961–2, pp. 107–19.

appears to have diverted his interests away from patent work. In 1830 and 1831 he apologised to Lord Cochrane for 'not being able to meet his Lordship [since] I have been out of town taking notes for another specification [and] ... fully employed on a series of specifications which have been accumulated'.[26] By 1835, though, this side of his business had become less active.[27]

Bennet Woodcroft was another engineer who became a patent agent. Born at Stockport in 1803, he was the son of a Manchester dyer and velvet finisher. After being apprenticed to a Failsworth weaver he took up the study of chemistry under Dalton. In 1826 he started manufacturing silk on his own account and in the following year took out his first patent.[28] Throughout the 1830s and 1840s his inventive activity diversified into calico printing and machine making, and during this period he experienced the difficulties of defending (and extending) his patents.[29] It was also a period when he became acquainted with that rising class of Manchester machine makers. He was a frequent guest at Medlock Bank, the home of William Fairbairn, and whilst writing for the *Workshop* periodical — which later failed — came into close contact with Whitworth, Nasmyth, Eaton and Hodgkinson. All these were patentees in their own right and, no doubt, frequently discussed their experiences of patents. Woodcroft, like many others, was incensed by the failure of the Patent Office to provide easy and cheap access to past specifications, and in the late 1830s he began to compile his own historical catalogue of patents. In 1843 he became the first Manchester patent agent, but within three years he had moved to London, where his impact was immediate. In 1848 he was appointed to the Chair of Machinery at University College, London,[30] and in 1849 he was called as a witness before the Select Committee examining the Signet and Privy Seal Offices. He gave evidence before the 1851 Select Committee and in 1852, after the Patent Law Amendment Act, was appointed Superintendent of the Specifications in the newly created Office of the Commissioners of Patents for Invention,[31] a post 'unexpected and unsolicited' but applauded by the *Manchester Guardian*.[32]

Carpmael, Farey and Woodcroft typify the general background of the early patent agents. Luke Herbert, William Newton junior, Thomas Gill, Joseph Robertson, Paul Hodges, William Spence were all of the same ilk.[33] In the 1840s a small number of legally trained men were, as A. V. Newton observed some years later,[34] beginning to offer their services to inventors, but never at any time before 1850 did they constitute an important part of a growing profession. In the main, patent agents acquired legal knowledge through experience and diligence, not through formal legal training — although there were a few exceptions.

For the early nineteenth century there is very little evidence of the services which patent agents offered. The growing volume of patent law literature, much of it written by patent agents themselves, is unhelpful and in 1850 one commentator was able to observe that the 'class of patent agents is by no means a well defined one'.[35] This remained the case until the 1870s, when A. V. Newton published in his *Patent Law and Practice* the first 'chapter in all the law books on the Patent Agents'.[36] The most explicit account of the services provided by the patent agent comes, however, from an article Newton wrote in 1882. Although it refers to some services which did not exist prior to 1852, it is the fullest description of what a patent agent did, and is therefore worth quoting at length:

> The popular notion of the duties and qualifications of a Patent Agent may be shortly stated as thus: He is fully informed in respect to the law and practice of patents as exemplified by the judgements of the courts and the decisions of the Law Officers. He is conversant with the several copyright Acts, and the mode of securing protection under the same. He is also familiar with Trades Marks Act and the practice under it, the Merchandise Marks Act, the General Acts relating to Gas and Water Companies, and is competent to advise respecting the establishment of manufacturing and trading companies under the limited liability acts. He is necessarily familiar with all the manufactures of this country, not to mention the various industrial exhibitions now so common in London, in the provinces and in foreign capitals. Moreover, he has traced the growth of the various branches of manufactures, from the earliest period of their inception to the present time, throughout the printed specifications of English Patents, all of which, now numbering 150,000 are open to his inspection in the Patent Office. He will advise as to the practicality of any mechanical contrivance or process submitted to him, as to the mercantile value of the same, how best the inventor may introduce his improvement to the trade, or to the general public; whether the event of finding some one inclined to take up the invention, a total or partial sale of the patent should be effected, or an exclusive or limited licence granted and on what terms ... [The Patent Agent will be] familiar with all the foreign and Colonial Patent Law [and] ... will also be able to advise how to prevent a manufacturer who owns a patent from intimidating the customers of a rival manufacturer by holding up the rival as an infringer ... His chief duties are to collect the inventor's ideas, to arrange them in a specification, which will eventually prevent any rival manufacturers from doing anything in the direction of the patent. If the invention is imperfect

at the time it is submitted to him, the Patent Agent will readily remove the difficulty ... his great experience giving him facilities which no ordinary inventor could be expected to possess.[37]

Newton was by no means sure that his account of the patent agent's duties were complete, although he was certain that the description was no exaggeration of what was expected. 'We have before us a high ideal which it should be our duty, as far as possible, to realize. It is useless,' he concluded, 'to take comfort in the fact that to reach this ideal is impossible.' That the ideal was rarely achieved should not disguise the fact that the patent agent's services were widely ranging and complex.

Apart from seeing patent applications through the hazards of the Patent Office,[38] agents assisted inventors in three ways: they prepared the specification, provided legal advice when patentees were considering litigation, and, lastly, acted as intermediaries between inventors, capitalists, innovators and other users of inventive output. The preparation of the specification was the most important part of their business.[39] In the late eighteenth and early nineteenth centuries, when judges were grappling with a new and developing law, varying interpretations of how good a particular specification was inevitably made it difficult for patentees to know where they stood. Few were competent enough to describe the nature and manner of the invention, and many were unable to distinguish between what was new and what was old. Most inventors were so impressed by their own originality that they rarely considered what the courts might think until it was too late.[40] Since most patents were set aside because of some fault in the specification, inventors came to rely on those who knew what the courts required. Naturally, some were capable of preparing their own specifications,[41] but even here they were insistently warned to take 'every possible care in drawing up your specification, as on this document will depend the security of your patent'.[42] In most cases the specifications were examined by the patent agent (hence their opposition to a Board of Examiners) and returned to the inventor if amendments were necessary. For the mass of inventors, though, patent agents were crucially important. Significantly, the improved quality of specifications in the 1830s and 1840s coincided with the changing attitudes of judges. In this sense, patent agents played a part in bettering the position of inventors generally.

The successful preparation of specifications[43] led to a strengthening of the relationship between patent agent and inventor. Inventors, now more assured that the courts would not throw their patents out, relied more and more on the patent agent's judgement. In turn, the patent agents began to expand into other areas. They helped inventors when their patents came before the courts, prepared briefs, and arranged to bring in the best witnesses. Carpmael, who reported every case from the 1820s onwards, frequently acted as a witness himself, and his judgement was widely respected, except perhaps on chemical matters.[44] Since many had built up their own catalogue of specifications on an industry-by-industry basis,[45] they had access to all recent developments, and were therefore able to inform clients of likely cases of infringement and put inventors in touch with each other. In some cases they were also able to dissuade inventors from attempting to patent inventions which had been patented by other clients.[46] Whether they succeeded in reducing the number of inventions patented is questionable, but at least they made inventors aware of the risks and costs involved. During the 1830s and 1840s ignorance of the law was no longer a reasonable excuse.

Patent agents were also in a position to advise inventors what to do with their inventions. Since they had close contact with a large number of manufacturers and capitalists, they often acted as brokers in the selling, licensing, assigning and financing of patents. William Carpmael and Joseph Robertson introduced many inventors to manufacturers looking out for ways of improving efficiency,[47] and John Farey reported to the 1835 Select Committee that he was frequently 'consulted on the propriety of seeing inventions after a patent has been obtained ... and the reason of that course having become very general with me of late years is [that] I have been chiefly employed by capitalists who have consulted me whether they should lay out their money in them'.[48] For many inventors, especially those without the requisite capital and those in the business of selling their inventive output, this service had obvious advantages. It saved them the problem of finding financial support and the trouble of hawking their inventions around the various firms, and protected them from exploitation by unscrupulous manufacturers. Working anonymously through a patent agent, inventors were shielded from face-to-face contact with men allegedly more aware of the 'cares and risks of commerce'. Users of inventions also benefited. Patent agents reduced search costs, and, since their judgement was well respected, manufacturers could be reasonably sure of investing their funds wisely. Patent agents would also advertise and promote the goods produced by inventors, contacting retailers, wholesalers and manufacturers to inform them of the latest developments. J. C. Robertson's letter to Samuel Moulton, the Bradford-on-Avon rubber manufacturer, is perhaps typical. 'Some tailors at the West End,' he wrote, 'have on *our* suggestion examined your water proof cloth. They think they could do an immense trade in coats and cut out all the others if the price would not prevent them ... They would,' he added, 'advertize largely and give your name of the cloth thus "Waterproof over coats, manufactured from Moultons Patent, India Rubber Cloth, soft, elastic and free from smell".'[49]

In a period when inventors were often stigmatised as schemers and where manufacturers were reluctant, for a variety of reasons, to take up new inventions, the patent agent fulfilled an important role in reducing the risk in this highly risky market. As Henry Dircks observed, 'An *if* attends every new invention: *if* it should not be superceded; *if* it has to incur the expense of a Chancery suit; *if* the patent is old; *if* the consumption increases and so on.'[50] In this world of 'ifs' the patent agent introduced some degree of assurance and trust by 'educating' both inventors and users of invention of the opportunities that existed, and thus made the market for inventive output more efficient.

Patent agents did not organise themselves into a professional body until 1882. During the early nineteenth century there was no institutional court of authority to control their activities or to set standards of competence. 'Any person,' Lloyd Wise complained in 1885,

> however incompetent or disreputable ... is at liberty to commence practice as a patent agent without passing on examination or possessing a licence or authority; whereas he would have to qualify before being allowed to practise as a barrister, solicitor ... physician, surgeon, dentist or chemist and would require a licence even to drive a cab, or an omnibus, act as omnibus conductor, sell tobacco, wine, beer, spirits, game, stamps, or fireworks or become a hawker or keep a carriage, a gun or a dog.[51]

Similar complaints were expressed in the 1830s and 1840s, and several attempted to discredit the growing number of patent agents.[52] The council of the Law Society reported in 1848 that the 'province of professional men' had been encroached 'by persons engaged as Agents in soliciting patents',[53] and in 1851 it recommended that the business of patents ought to be confined to solicitors, because patent agents were 'without legal education and regular qualification'.[54] Richard Prosser suggested that the wholesale removal of patent agents was desirable 'for the sake of promoting morality',[55] and Christian Allhusen, president of the Newcastle and Gateshead Chamber of

Commerce, went so far as to claim that patent agents were the only 'class to benefit from patents'.[56] A number of inventors felt much the same, although with little justification.

Patent agents were accused of many things. Benjamin Fothergill claimed he knew of one agent who had prepared three specifications within a period of eighteen months for exactly the same process,[57] and Thomas Fowler maintained that Benjamin Rotch had advised one London firm to 'infringe his patent with impunity' because he had ignored some of Rotch's suggestions.[58] Paul Hodge also claimed that some patent agents frequently used caveats as 'standing dishes' for the purpose of having 'individuals to come to oppose, in order to get a fee for buying off their opposition and for nothing else'.[59] J. Prince in an outraged letter to Lord Brougham expressed his belief that in the

> last twenty or thirty years the policy of the law had been defeated by a junta of patent agents who have made it their study and their livelihood to ascertain by what devices the invader of a patent can be upheld, and this systematic injustice has at length grown up under the public neglect, to such a pitch of enormity, and is carried on with so much cunning, as well as audacity, that I cannot forbear from imploring your lordship to give full effect on th[is] important subject.[60]

There is little doubt that some patent agents abused their position of confidence and that some made genuine mistakes, but there is little evidence to suggest that there was any substance in the allegations. In fact, inventors were not always as innocent as they claimed.

Patent agents taking out patents in their own name provided another source of complaint. Between 1816 and 1852 some 537 patents were taken out by agents.[61] Here again, though, there is no evidence to suggest that this increased the alleged 'numerous frauds now committed against inventors'.[62] In fact patent agents took out patents in their own name for three legitimate reasons. Firstly, if English inventors had used their own name for a patent taken out in Europe this precluded them from using their name for a patent taken out in another country. 'A patent taken out in your name in France must afterwards,' so Charles Albert told Stephen Langton, 'be taken out in another ... or the patent is invalid'.[63] Albert suggested Newton and Co, but Langton stuck with Moses Poole, who, in 1827, reminded him that the 'patent although taken in my name belongs exclusively to you; which I readily admit and am willing to assign it to you as soon as you require'.[64] Secondly, patent agents took out patents in their own name because a number of inventions were communicated to them by foreign inventors. A. V. Newton estimated that half the American inventions protected by British patents passed through his father's firm.[65] How many other foreign inventors relied upon patent agents remains unknown, but in general, as T. Leonard observed, patents for 'foreign inventions are obtained by a class of men called ... patent jobbers'.[66] Finally, some patent agents were themselves inventors and at various times manufactured goods on their own account.

The fact that many of these patents were taken out by agents who worked in the Patent Office supplied further grounds for concern. Some vehemently opposed the idea that officials should act as agents, and one letter writer in *The Times* went so far as to suggest that 'a large penalty of imprisonment should be awarded to an officer who becomes an agent for taking out patents'.[67] This was clearly a harsh judgement but one which seemed to appeal to the *Mechanics Magazine*, and for obvious reasons:[68] Robertson was not so well placed as Poole, Newton and Berry. In 1829 Benjamin Rotch claimed he knew of a case where £100 was given to hurry the patent application through the Patent Office, but whether Poole, Newton or Berry was involved again remains unknown.[69] Their privileged status, however, made them an obvious target for criticism, and their rigid opposition to patent reform did nothing to allay the innuendoes and rumours.

The growth of patent agents in the early nineteenth century demonstrates that inventors and entrepreneurs depended (rather more than is usually supposed) upon expert advice in the management of their affairs. The market for inventive output was exceptionally risky and made all the more so by the judgements of the courts. This created an opportunity for men to offer their skills as consultants in a very specialised and technically complicated area. They advised on almost every aspect of the invention industry and generally (despite some unjustified criticism) sought to relieve inventors from making costly mistakes. Indeed, if the emergence of the service sector during the industrial revolution marked a 'significant break with the past',[70] then the growth of the patent agent was one of the most important developments in the invention market.

**Notes**

1 *Newtons Journal*, XXXVI, Conjoined Series, 1850, pp. 321–4.
2 Coryton, *op. cit.*, p. 147.
3 See R. M. Hartwell, *The Industrial Revolution and Economic Growth*, 1971, pp. 201–25; and 'The service revolution', in Cipolla, ed., *The Fontana Economic History of Europe*, 1973, III, pp. 358–96.

## Patent agents

4 Many of the early patent agent firms still exist today, but none was able to provide any records.
5 Robinson suggests the late eighteenth century saw the emergence of the patent agent as a legal expert but offers no evidence to support the assertion, *op. cit.*, 1971, p. 116.
6 Boulton and Watt MSS, *Thoughts upon Patents or exclusive Privileges for new Inventions*, Box 21, B.R.L., Robinson and Musson, *op. cit.*, 1969, p. 217.
7 R. Robson, *The Attorney in Eighteenth Century England*, Cambridge, 1959, p. 164; 'The diary of Samuel Taylor', MS/925/775, M.R.L.; Wedgwood MSS, Wedgwood to Bentley, 22.10.1770, John Rylands University Library of Manchester.
8 F. M. L. Thompson, *Chartered Surveyors: the Growth of a Profession*, 1968, p. 64.
9 *Report of the Committee on the Signet and Privy Seal Offices*, Parl. Papers, XXII, 1849, p. 520.
10 Poole was appointed by Sir Samuel Sheppard but this was not confirmed until Lord Lyndhurst did so in 1826, *ibid.*, p. 578.
11 *Select Committee on the Design Act Extension Bill*, Parl. Papers, XVIII, 1851, p. 693.
12 *Select Committee on Patents*, Parl. Papers, III, 1829, p. 16.
13 W. Newton Jnr., *In Memoriam: being a memoir of the late W. Newton*, 1861, p. 5.
14 See the comments of John Henry Johnson, the first president of the Institute of Patent Agents, formed in 1882. *Transactions of the Institute of Patent Agents*, I, 1882–83, p. 39.
15 *Dictionary of National Biography*.
16 *Select Committee on Patents*, Parl. Papers, XI, 1872, p. 65.
17 *Select Committee on the Signet and Privy Seal Offices*, Parl. Papers, XXII, 1849, p. 521; W. R. Forwell, ed., *Chartered Institute of Patent Agents: Informals: Collected Papers*, 1969, p. 1.
18 *Select Committee on the Signet and Privy Seal Offices*, Parl. Papers, XXII, 1849, p. 481.
19 *Select Committee on the Design Extension Bill*, Parl. Papers, XVIII, 1851, p. 647.
20 Webster's *Reports* I, pp. 377–93; Carpmael's *Reports* II, p. 649.
21 *Select Committee on the Signet and Privy Seal Office*, Parl Papers, XXII, 1849, p. 491.
22 R. B. Prosser, *Birmingham Inventors*, 1881, p. 66.
23 *Minutes of the Proceedings of the Institution of Civil Engineers* XI, 1852, pp. 100–2. Between 1821 and 1823 he set up a lace factory but the affair ended in failure.
24 Boden MSS, Bo. 24, 1.11.1817, Nottingham University Library.
25 W. Felkin, *History of the Machine-wrought Hosiery and Lace Manufacture*, 1967 ed., p. 204.
26 Dundonald MSS, Farey to Cochrane, 20.9.1831, GD/233/2; see also 1.9.1830, 29.9.1830, 29.9.1830, 7.3.1830, 28.4.1831, GD/233/31, Scottish Records Office (hereafter S.R.O.).
27 *Lords Select Committee on Patent Amendment Bill*, 1835, H.L.R.O.

28 Woodcroft had thirteen patents in his name but took some of these out for other inventors.
29 For the extensions of his patent see *The Commissioners of Patents Journal*, 1859, pp. 446–63.
30 *Alphabetical Index of Patentees*, 1969 ed. Woodcroft did not suit the role of academic and so resigned in 1851.
31 This brief account of Woodcroft is based on *D.N.B.* and obituaries in the *Manchester Guardian*, 11.2.1879; *Times*, 14.2.1879; *Journal of the Society of Arts*, 21.2.1879; *The Engineer*, 14.2.1879 (written by R. B. Prosser).
32 *Manchester Guardian*, 22.12.1852.
33 *Select Committee on Patents*, Parl. Papers, XVIII, 1851, p. 309. *Select Committee on Patents*, Parl. Papers, XXIX, 1864, p. 381.
34 A. V. Newton, *Patent Law and Practice*, 1879, p. 78. Later in the nineteenth century a growing number of patent agents were recruited from the legal profession.
35 T. Turner, *Counsel to Inventors*, 1850, p. 98.
36 A. V. Newton, *op. cit.*, p. 77.
37 A. V. Newton, 'The patent agent's profession', *Transactions of the Institute of Patent Agents*, I, 1882–83, pp. 159–60.
38 See W. Carpmael, *The Law of Patents for Invention*, 1832, p. 20; *Mechanics Magazine*, XXX, 1839, p. 436; *Select Committee on Patents*, Parl. Papers, XVIII, 1851, p. 314. Some, like J. C. Robertson, issued to their clients 'Instructions to intending patentees', *The Act to Consolidate and Amend the Laws Relating to Copyright and Design*, 1842, p. 23.
39 See, for example, Holden MSS, Newton and Sons to Holden, 20.10.1843, Box VI, Bundle 4, Brotherton Library, Leeds (hereafter B.L.).
40 *Mechanics Magazine*, V, 1859, p. 69.
41 See, for example, *Lords Select Committee on the Patent Amendment Bill*, 1835, H.L.R.O.; *Select Committee on Patents*, Parl. Papers, XVIII, 1851, pp. 187, 203; Raistrick MSS, Poole to Raistrick, 4.7.1815. AL/340/12, London University Library; H. I. Donkin, *Bryan Donkin and Co: Notes of History of an Engineering Firm during the Last Century, 1803–1903*, 1912.
42 Dundonald MSS, Poole to Lord Dundonald, 6.12.1843, GD/233/6–12/61 S.R.O.
43 Both Carpmael and Farey claim they lost only one patent because of a faulty specification. *Select Committee on the Design Extension Act*, Parl. Papers, XVIII, 1851, p. 671; Prosser, *op. cit.*, p. 66.
44 *Select Committee on the Design Extension Act*, Parl. Papers, XVIII, 1851, p. 693.
45 Society of Arts MSS, D7/263; Dundonald MSS, Poole to Dundonald, Aug. 1843, GD 233/6–12/61 Box 6, S.R.O.
46 *Select Committee on the Design Extension Act*, Parl. Papers, XVIII, 1851. Carpmael claimed he advised six out of every seven inventors not to take a patent out. See also *Select Committee on Patents*, Parl. Papers, XXIX, 1864, p. 360; James Young MSS, J. C. Robertson to Young, 25.6.1850, MS/fo/2, Andersonian Library, University of

Strathclyde; Spencer Moulton MSS, Newton and Sons to S. Moulton, 1.4.1850, 3.4.1850, 22.7.1850, 19.8.1851; Holden MSS, Newton and Sons to Holden, 30.6.1843, B.L.; Chadwick MSS, Poole to Chadwick, 2.9.1848, U.C.L.; Marshall MSS, Miles Berry to J.G. Marshall, 21.7.1836, E/17/31, B.L.
47 *Select Committee on Patents*, Parl. Papers, XVIII, 1851, p. 49; *Select Committee on Arts and Manufactures,* Parl. Papers, V, 1835, pp. 501–2; Society of Arts MSS, R. A. Brooman to F. Wishaw, 15.5.1845, Bl/401.
48 *Lords Select Committee on the Patent Amendment Bill*, 1835, H.L.R.O.
49 Spencer Moulton MSS, Robertson to Moulton, 27.11.1850. See also P. L. Payne, *Rubber and Railways in the Nineteenth Century*, 1961.
50 Dircks, op. cit., 1867, p. 80.
51 Lloyd Wise, 'On patent agents: their profession considered as a necessity, with suggestions for reform', *Transactions of the Institute of Patent Agents*, IV, 1885–6, p. 79. For the development of the profession in the late nineteenth century see A. M. Carr-Saunders and P. A. Wilson, *The Professions*, Oxford, 1933, and *Select Committee on Patent Agents Bill*, Parl. Papers, XIV, 1894.
52 See Brougham MSS, [29,248], T. Webster to Brougham, 29.5.1844, U.C.L.; Hindmarch, op. cit., pp. 503–5.
53 Report of the Council to Annual General Meeting, Minute Books of the Law Society, 30.5.1848, p. 23.
54 Report of the Council to Annual General Meeting, Minute Books of the Law Society, 19.6.1851, p. 16.
55 *Select Committee on Patents*, Parl. Papers, XVIII, 1851, p. 314.
56 *Select Committee on Patents*, Parl. Papers, XI, 1872, p. 103.
57 *Select Committee on Patents*, Parl. Papers, III, 1829, p. 204.
58 Brougham MSS, [46,624], Fowler to Brougham, 22.1.1834, U.C.L.
59 *Select Committee on Patents*, Parl. Papers, XVIII, 1851, p. 320.
60 Brougham MSS, [47,370], Prince to Brougham, 1834, U.C.L.
61 Of these, 192 were taken out by Newton and Co.: Moses Poole registered 106 patents between 1817 and 1852, Pierre Fontainmoreau fifty, Miles Berry forty-one, R. A. Brooman thirty-nine, L. Herbert eighteen, C. Cowper sixteen, J. Murdoch fifteen, J. C. Haddon and B. Woodcroft thirteen, J. C. Robertson eleven, Joshua Bates eight, W. Ritchie seven, D. Barlow, J. Bethall and B. Rotch five, whilst the remainder were taken out by W. Baddeley, F. Campin, W. Carpmael, H. Dircks, C. Dod, J. Gedge and T. Gill.
62 Brougham MSS, [9,711], Sir F. C. Knowles to Brougham, 29.4.1851, U.C.L.
63 Langton MSS, Albert to Langton, 6.12.1825, 3/1/3, Lincolnshire County Record Office (hereafter L.R.O.).
64 Langton MSS, Poole to Langton, 21.9.1827, 3/1, L.R.O.; see also Dundonald MSS, Poole to Lord Cochrane, GD/233/1/38, I J, S.R.O.
65 *Select Committee on Patents*, Parl. Papers, XVIII, 1851, p. 415, Moulton MSS, Newton to Moulton, 1.4.1850.
66 P.R.O. Granville Papers, 30/19/23/15/63, T. Leonard to Lord Granville, 4.12.1852.

67 *Times*, 13.2.1835. See also A. Rosser's evidence taken before the *Lords Select Committee on Patent Amendment Bill*, 1835, H.L.R.O.
68 *Mechanics Magazine*, XXIV, 1835, p. 64.
69 *Select Committee on Patents*, Parl. Papers, II, 1829, p. 117.
70 Hartwell, *op. cit.*, 1973, p. 362.

# PART II

Patents and inventive activity

# 6

# Invention and inventive activity

Economic historians have considered invention as a vital part of the story which seeks to explain the English industrial revolution. Writing as early as the 1850s, James, the historian of the worsted industry, noted that the 'whole world is indebted to ... [English] ... inventions', and that as 'her just reward [England] has reaped, beyond all comparison, the largest share of the golden results which have flowed from them'.[1] In the 1880s Toynbee saw the substitution of the factory for the domestic system as essentially the 'consequence of the mechanical discoveries of the time'.[2] Cunningham started Part two of the *Growth of English Industry and Commerce* by referring to Arkwright's invention and its connection with the 'commencement of a new era in the economic history, not only of England, but the whole world' (p. 1). Clapham, on the other hand, was less enthusiastic, and following his observation that 'no single British industry has passed through a complete technical revolution before 1830',[3] gave less weight to the role of invention in the industrial revolution. This remained largely the case until economists discovered the 'residual'[4] in the 1950s and 1960s, since which time economic historians have again sought to emphasise invention and technological change. Lilley, for example, characterises the industrial revolution as 'a more or less rapid transition from the domestic workshop to factory production, and from a *manu*facture in the literal sense to *machine*facture', where 'a spate of revolutionary inventions entirely transformed the technological scene'.[5] In the *Unbound Prometheus* this theme is embellished in style. 'In the eighteenth century,' Landes writes, 'a series of inventions transformed the manufacture of cotton in England and gave rise to a new mode of production — the factory system', and 'these improvements constitute the Industrial Revolution'.[6] For Rostow 'what distinguished Britain from the rest as the eighteenth century wore on was the scale of inventive effort that went into the breaking of crucial technical bottlenecks'.[7]

Invention thus plays an important role in the explanation of the industrial revolution. Inventions lead (although not always) to innovations, innovations then become diffused, and this in turn leads to changes in the forms of production and to increased efficiency. No one would quibble with the logic of the argument, even though some may hold that there are other causes of the industrial revolution. What remains something of a mystery, though, is the almost total neglect on the part of economic historians to analyse invention and inventive activity.[8] Most, it seems, concentrate on the determinants of innovation and diffusion without saying very much about the original invention, except perhaps from a technical or scientific perspective.[9] Naturally, the 'spectacular inventions in steam, waterframe, spinning jenny, coke, puddling and the like occupy the front and centre of the stage of the Industrial Revolution',[10] but, for all this, there has not been any serious investigation of what causes the production of inventive output.

This, and the following chapter, will provide a largely economic explanation of inventive activity and the production of inventive output. It will also be argued that an *infant* invention industry emerged during the industrial revolution. The first section of this chapter will briefly survey the recent literature. In the second section four criteria for evaluating invention as an economic activity will be suggested. Subsequent sections will examine certain aspects of the invention industry in the light of three of these criteria. The trade in invention will be discussed in the next chapter.

The notion of the inventor as 'economic man' has not been readily accepted by economists or economic historians.[11] In most cases it has been rejected in favour of a more eclectic approach, which gives expression to a variety of non-economic determinants ranging from the instinct of contrivance, the inborn and irresistible impulse to acquire knowledge, tinkering, intuition and accident, to the rather more grandiose pursuit of fame and the betterment of mankind.[12] Profit sometimes does have a role (usually assumed rather than demonstrated),[13] but rarely is invention seen as an economic activity in its own right. Nor, consequently, is there an explanation of the market that this kind of activity necessarily creates.[14] Even in Schumpeter's work, where one might have expected to find an analysis of inventive activity and the production of invention, a blank is drawn. By asserting that invention 'produces of itself no economically relevant effect at all',[15] he conceives the inventor in strictly non-economic terms, and thus repeats the view that had previously been given extended currency by the works of Pigou and Stamp, both of whom saw invention largely as an autonomous, spontaneous and exogenous variable, 'touched by no economic spring'.[16]

The first notable reaction against these views came from the sociological school of Chicago, which dismissed the 'heroic' conception of invention implied by the supporters of the 'exogenous case'. Through the works of Gilfillan and Ogburn this school extolled the impersonal forces of social change. Here individual inventors are mere tools of 'cosmic forces', and inventions inevitably respond to social needs and tastes through the unceasing accretion of inventive detail. Science and scientific knowledge are vitally important aspects of this explanation of inventive activity, but by neglecting to consider either what motivates inventors or the impact of economic forces, demand factors were not examined.[17]

This was done by Schmookler in his *Invention and Economic Growth*.[18] By demonstrating that variations in inventive activity over time and between industries are related to variations in investment goods and the volume of sales of capital goods, Schmookler argues for a demand-induced explanation of inventive activity. This is a modern variant of the adage 'necessity is the mother of invention', where inventive activity responds to economic needs. In short, Schmookler treats inventive activity as an endogenous variable, which can therefore be analysed in exactly the same way as any other form of economic activity. Inventive activity is an economic phenomenon.

Schmookler's thesis has been criticised on a number of counts. Jewkes *et al.* believe that Schmookler 'tends to over-emphasise the importance of demand', because many inventions developed out of a sense of craftsmanship and because inventors 'often had no idea to what purpose their inventions can or might be put'. Moreover, although many inventors were years ahead of their time they continued to perfect their inventions despite knowing that the achievement of perfection depended upon simultaneous technical progress elsewhere. For both of these reasons Jewkes *et al.* suggest that 'economic motives are secondary'.[19]

Rosenberg takes a similar, if very much more sympathetic view, for quite different reasons. His major criticism is that Schmookler pays too little attention to the role of science and knowledge as a determinant of inventive activity. Demand, he argues, cannot possibly explain all inventions, because too many human wants have gone unsatisfied despite strong evidence of 'well-established demand'.

The supply of inventive output is therefore not perfectly elastic. Moreover, demand has a limited explanatory value because it is not defined independently of the evidence that demand was itself satisfied. 'In the absence of a reasonably clear, independent specification of the composition of demand, one can never demonstrate either that important components of demand have gone unsatisfied or that supply side factors [such as science and scientific discovery] played an important role in laying down the time pattern of inventive activity'.[20]

Finally, Schmookler's analysis seeks only to explain the economic determinants of invention in aggregate. He is not concerned with the motives of individual or independent inventors.[21] This raises two problems. Firstly, his analysis does not preclude independent inventors from being motivated by non-economic factors.[22] Secondly, there is no assessment of the market relationship between inventors and the *users* of inventive output. Schmookler is largely concerned with the 'captive' inventor and with firms who buy their own inventions. Consequently, the expected profitability which explains inventive activity is related to entrepreneurial profit rather than to the profit the inventor as an individual might expect to earn. For historical analysis this is not very appropriate, largely because in the late eighteenth and nineteenth centuries the private independent inventor was generally a major source of technical knowledge.

These criticisms of Schmookler's thesis indicate that economists and economic historians are not entirely convinced that invention is an economic activity.[23] Some have simply been reluctant to accept the notion, and perhaps for two reasons. Firstly, there is ample evidence to show that failure and bankruptcy frequently accompanied invention — indeed, the nineteenth century is littered with the corpses of failed inventors. This, understandably, has led many to consider the inventor as a species of mankind not overendowed with business acumen. But, as with the romance of success, the romance of failure can be misleading. Bankruptcy, as Adam Smith wisely noted, 'is perhaps the greatest and most humiliating calamity which can befall an innocent man. The greater part of men therefore are sufficiently careful to avoid it. Some, indeed, do not avoid it; as some do not avoid the gallows.'[24] That many inventors failed to avoid the 'gallows' can be explained largely by the fact that bankruptcies, to quote Smith again, 'are most frequent in the most hazardous trades'.[25] Inventors worked in a high-risk market, where there were huge discrepancies between invention and innovation and where the 'success of one inventor often [brought] loss on hundreds of other inventors'.[26]

In these circumstances, it is not really surprising to find that many inventors experienced the ignominy of the financial 'gallows'. It is important to note, though, that they did not have a monopoly of failure. Nor can failure logically be employed as an argument to show that inventors were not concerned with profits: the failure to make profits is no indication of motive or intent.

The image of the unbusinesslike inventor is given further credence by the fact that many were demonstrably out of touch with commercial reality when it came to *working* their own inventions. Yet such failures are an indictment of their abilities as innovators, not as inventors, and these are two quite separate economic activities. Inventors may well have been unable to make money as captains of industry, but this is not to say that they were equally incompetent when it came to the business of producing and selling inventive output. 'The truth,' one contemporary wrote, 'is that inventors and perfectors would have a far greater stimulus to exertion when sure of compensation without the necessity of becoming either merchants or manufacturers. They would labour in the vocation for which they are fitted without feelings of anxiety, lest they should be driven into vocations for which they are entirely unfitted.'[27]

The second reason why inventors have rarely been treated as economic men is that many inventions were 'ahead of their time'. Musson argues this point with force when referring to a number of eighteenth-century improvements. 'If these inventions were simply products of pressing economic and social forces, why was there such a long time lag before their widespread application? Surely, if they were ... economically "determined", "inevitable", and "necessary" they should have been brought into widespread use immediately.'[28] The thrust of this argument is compelling (especially its critique of deterministic explanations of invention), but it seems to be based on a number of interrelated assumptions which may not be generally valid. Firstly, it assumes that inventors always respond to demand and ignores the fact that they may *create* a demand for their own output in the expectations of making profits. Secondly, it assumes that the *producers* and *users* of invention have an identical perception of what is needed to reduce costs (or to introduce a new product), whereas it may be that inventors see well in advance of the user what is needed but cannot provide convincing proof for some time. Thirdly, it assumes that there is only one invention which is known that will solve the problem which a user may identify, when in fact there may be several inferior competing inventions which are tried first.[29] Lastly, it assumes there are no gains for innovators who *wait*. If the rate of technological change rapidly makes today's invention obsolete, it may pay to postpone making a decision until tomorrow's invention has appeared. Conversely, if the rate of technological change is relatively slow, it may pay to adopt today's technology rather than wait until tomorrow. Expectations concerning the rate of technological change will thus determine the speed at which inventions are taken up, and this will, in part, occur independently of the nature of individual inventions.[30] In short, the 'ahead of their time' argument is a critique of economic determinism, not a critique of economic behaviour, and it does not preclude inventors from being motivated by profit *ex ante*. In the market for inventive output demand and supply interact unequally over time, and user demand is not always the main determinant. Like any other producers, inventors will attempt to create a demand for their output in the expectation of making profits.

Clearly, there is very little agreement about what causes inventors to invent. Nor is the evidence helpful in sorting out this problem. Inventors were not great letter writers and few wrote books. Many were sanguine enough to let their inventions speak for themselves.[31] The lack of detailed evidence on a wide cross-section of inventors makes it difficult to test the proposition that inventors generally invented in the expectation of making a profit. Nevertheless by judicious use of patent statistics and information concerning the sale of patents it can be shown that a considerable number of inventors were indeed economic men operating in what might be termed an invention industry. There would seem to be at least four criteria that would have to be satisfied if this proposition is to hold:

1. That the bulk of inventors should go to the trouble and expense of obtaining patent protection for their ideas.
2. That a considerable proportion of patented inventions should be taken out by quasi-professional inventors, that is, inventors holding several or numerous patents.
3. That quasi-professional inventors, working in a high-risk industry, should diversify their inventive portfolio by inventing in a number of different areas or industries.
4. That a vigorous trade in (patented) inventions should exist.

Each criterion will be examined in turn.

The number of patents taken out by inventors during the industrial revolution is the first indication that inventors were primarily concerned with profits. Patents, as Babbage was not alone in observing, were 'devised to assist and reward those who have chosen the line of pecuniary profit'.[32] 'In reference to trade,' another contemporary remarked, 'there is no known mode by which a great invention can be made a commensurate source of profit, except that which operates through possession of a temporary monopoly given by a patent.'[33] According to Bessemer, 'there can be no doubt of the fact that the security offered by the patent law to persons who expend large sums of money and valuable time in pursuing novel invention, results in many new and important improvements in our manufactures, which otherwise it would be sheer madness for men to waste their energy and their money in attempting'.[34] And no less an inventor than Watt 'feared that an engineer's life without patent was not worthwhile'.[35] In a letter to Lord Loughborough Watt revealed why inventors invent. Since this is an almost unique record of what one particular inventor thought it is worth quoting extensively:

> There seems to be only three motives that can excite a man to make improvements in the arts, the desire of doing good to society, the desire of fame, and the hope of increasing his private fortune. The two first, when unmixed with the latter, ought only to actuate men who are already independent or have a competency of the goods of fortune; for it would be reckoned folly in any man in circumscribed circumstances to devote his time and money solely to the public good or to the pursuit of the bubble[?] reputation; but when the three motives are united they must prove the strongest stimulus which can act upon the human mind, and they can make it to struggle with anxiety, the unremitting attention and the frequent disappointment and the labour and expense which infallibly attend every attempt at improvement in the arts ... In consequence of the security which we imagined these exclusive [patent] privileges gave, we have for many years devoted our time and money to the bringing the invention to perfection and though the publick has been already considerably benefited thereby; yet we have not hitherto acquired such sums of money by it as should make it be esteemed an enviable concern were the profits to terminate at this juncture. We are now bringing steam engines into use as a moving power for mills of many kinds which cannot fail of being a great advantage to the manufacturers of this country; but if our right to our patent should be taken away, or rendered illusive, we must drop any further pursuits of that scheme and apply ourselves to other businesses where our property can be more effectually guarded.[36]

Patents, therefore, were considered essential for inventors who wished to make profits. They provided an important, if varying, degree of security and protection in an otherwise very risky market, and enabled inventors to appropriate a return to their ingenuity and effort. Fame and the public good were secondary considerations.

The crucial question is, what proportion of inventions were patented? It is well known that patents do not reflect the quantity (or quality) of inventive activity, and how many inventions were unpatented and whether the ratio of patents to invention was constant over time remain a subject of disagreement.[37] Boehm and Silberston argue that the first eighty years of the nineteenth century constitute the 'age of patentless invention',[38] whereas Jewkes *et al.* see the nineteenth century as a period when the patent system 'flourished most vigorously'.[39] Both agree, though, that patent activity increased remarkably over the period.

To estimate the extent to which inventions were patented would require a monumental search and a good deal of fortune. There are three arguments, however, which have been advanced to show that patents do not reflect the volume of inventive activity: the cost of patenting, secrecy and changes in the standards for patentability.

Most (although not all) contemporaries thought patent costs excessive. Both the 1829 and 1851 Select Committees provide ample evidence to show that 'under the present system, we have not half as many inventors as we should have if we had cheap patents'.[40] This argument should be treated with caution, for two reasons. Firstly, most complaints concerning costs were part of the reform rhetoric, and a number of inventors wanted cheaper patents not because the present system prevented them from patenting but because they saw no good reason why the costs were as high as they were. Secondly, very few inventors actually paid as much as £400. Most, as has been shown, took out patents only for England and these cost on average £100–£120.[41] Moreover, many patentees were able to share the cost of patenting with their employers, partners and other inventors. Few were seriously deterred from 'prosecuting a favourable and useful pursuit, from the consideration that he could not raise the money to purchase a patent'.[42] Costs, therefore, were unlikely to have deterred *many* inventors from patenting in *England*, especially since this was effectively the only means of protection.

Secrecy was another factor affecting the ratio of patents to invention, although the nature of the problem makes it almost impossible to assess how widely it was practised. Inventions in chemical and allied industries were, allegedly, easy to keep secret,[43] but in other industries it seems to have been more difficult. Some inventors and many more manufacturers would place their employees under a bond, although this does not appear to have been manifestly successful. Richard Roberts did not think there was 'much secret trade ... but I know this, that no trade can be kept secret long; a quart of ale will do wonders in that way'.[44] Richard Arkwright had thought much the same: 'we might swear [the hands] as we pleased, but if any body would give them a penny more, they would divulge it'.[45] Many inventors feared the costs of successful industrial espionage and thought patents less risky than having their inventions pirated without any legal protection.[46] Much as inventors wanted to keep their inventions secret (even when patented), it seems that keeping secrets was not costless. Technology during the industrial revolution was not sophisticated enough to prevent imitation, and skilled labour proved to be fairly mobile when the inducements were sufficiently attractive. In short, the costs of secrecy probably exceeded the costs of patenting.

Changes in the standards employed by the Patent Office officials to test whether an invention merited patent protection have provoked another criticism of using patents as a surrogate for invention. In his study of the American antebellum patent system R. C. Post claims that this alone is reason enough to 'banish the assumption that [patents] provide any but the most tenuous basis for generalising about the trends in the economy or about invention'.[47] Unlike the American system, however, the British Patent Office did not alter its standards throughout the industrial revolution. Except for the caveat system (which allowed inventors to oppose competing patent applications), there was no examination of patents to test for originality. In Britain the Patent Office operated a system of registration. If inventors were prepared to pay the fees, patents were granted as a matter of course. The process of granting patents may well have been cumbersome, but it was not testing. Consequently, it seems likely that a large proportion of unimportant and imitative inventions were patented.[48]

The arguments against using patents to reflect the quantity of invention during the industrial revolution are thus far from convincing, and they become even less so if the informed opinions of contemporary commentators are to be believed. One anonymous observer writing in the 1790s argued that 'at present ... it seems a fashion to take a patent for almost everything'.[49] In 1824 Joseph Hume claimed that 'there was scarcely a single mechanical improvement that was not immediately protected by a patent'.[50] Twenty years later the Society of Arts reluctantly admitted that 'it is notorious that the most important inventions of the present day are secured by patents'.[51] In the evidence collected by the 1851 Select Committee similar views were expressed. J. L. Ricardo, an arch-opponent of patents, stated that 'everything is now done by machines, and unfortunately nearly everything under a patent', and Bennet Woodcroft, who had a vast experience of patents and inventions, argued that he did not 'know any good invention which [had] not filtered through the patent laws'.[52] Inventors, the *Mechanics Magazine* noted,

> especially are prone — perhaps beyond all others — to press forward, heedless of all but the future. Once visited by an idea new to himself, and full of seeming promise, the inventive enthusiast straitway [*sic*] demands ... patent protection ... and at once rushes forward to squander his capital, or to make his fortune, as the case maybe ... So little does he doubt the originality of that which, in his own eyes, glitters with novel brightness.[53]

None would have agreed with the notion that the nineteenth century was the 'age of patentless inventions'. Knowledgeable contemporaries believed that almost all the important inventions were patented. The perennial complaint that too many patents were useless suggests that there were a great many more besides. In 1776 one writer broke into verse:

> The time may come when nothing will succeed
> But what a previous Patent hath decreed
> And we must open on some future day
> The door of Nature with a patent key.[54]

In 1846 the following, if less poetic, lines were written:

> Patents are intimately connected with the Arts and Science of the country, their increased numbers attest the progress we have made — the vessel is propelled through the ocean by a Patent Screw, whose motive power is a Patent Engine — the vivid Railway train is whirled along by a Patent Locomotive — the thoughts of distant men are, with the speed of lightning, communicated by the Patent Electric Telegraph — we are carried by Patent vehicles, our habiliments are spun by Patent Loom, our hearths derive their cheerful aspect from Patent Stoves; whilst our lucubrations are assisted by the Patent Lamp. The ponderous engine, and the ladies' parasol — the blasting of iron, and the manufacture of bread, are alike Patent.[55]

In short, although patents and invention are not synonymous, a large number of inventions were patented during the industrial revolution,[56] and the patents were taken out because inventors anticipated making profits. The first criterion would, therefore, seem to be satisfied.

The second criterion which has to be met if inventors are to be regarded as economic men is that a large proportion of all patented inventions are taken out by multiple and quasi-professional inventors.

# Invention and inventive activity 113

The crucial factor here is that inventors should devote a considerable amount of time producing inventive output and hold several or numerous patents. Inventing would, as a result, tend to become a specialised economic activity, with a large core of producers generating the bulk of the inventive output. This is not to suggest that inventors did not spend the greater part of their lives working at other trades, or that single patent owners were not economic men. Multiple patenting would indicate, though, that making profits was the primary reason for inventing.

An analysis of patent statistics reveals that a significant proportion of patented inventions were produced by multiple patentees. Moreover, the proportion rose steadily as the industrial revolution progressed, increasing from 28·3% of all patents granted in the decade 1751–60 to 60·0% in the decade 1831–40. The proportion falls for the decade 1841–50, but this almost certainly reflects the fact that multiple patentees who took out only one patent prior to 1852 are excluded from the calculation,[57] which is based on data terminating in 1852. What is particularly striking, however, is the growth in the proportion of patents taken out by inventors who owned four or more patents, and who might reasonably be regarded as quasi-professional inventors. The proportion increases from 11·1% in the decade 1751–60 to 33·3% in the decade 1831–40 (see table 5). Truly 'heroic' inventors, holding eight or more patents, appear to have been at their most important between 1781 and 1830, when they consistently accounted for two-thirds of all patents taken out by those holding four or more patents, and around one fifth of total patents sealed.

The significant proportion of patent inventions taken out by multiple and quasi-professional inventors seems to satisfy the second criterion. Quantity, of course, is rather different from quality, and sceptics might argue that a number of these prolific inventors, encouraged by early or occasional success, tended to be overoptimistic when it came to patenting their ideas. This, however, was a general characteristic of patentees, and there is no evidence to suggest that the quality of inventions produced by quasi-professional inventors was inferior to those patented by others. In fact, multiple inventors produced many of the major inventions of the industrial revolution.[58]

The third criterion is that inventors should diversify their invention portfolio by inventing in a number of different industries. The reasoning here is that economic agents attempt to maximise profits within a risk constraint, and generally do so by holding different classes of

*Table 5  Percentage of total patents sealed 1751–1850 taken out by multiple patentees*

|  | \% attributable to patentees holding |  |  |  | All multiple patentees | Total No. of patents sealed |
|---|---|---|---|---|---|---|
|  | 2 patents | 3 patents | 4+ patents | 8+ patents |  |  |
| 1751–60 | 15·2 | 2·0 | 11·1 | – | 28·3 | 99 |
| 1761–70 | 21·3 | 8·1 | 9·5 | 4·7 | 38·9 | 221 |
| 1771–80 | 15·5 | 7·7 | 10·4 | 3·0 | 33·7 | 297 |
| 1781–90 | 21·7 | 3·7 | 16·8 | 10·6 | 42·2 | 512 |
| 1791–1800 | 19·1 | 8·3 | 21·2 | 16·8 | 48·6 | 675 |
| 1801–10 | 17·2 | 8·3 | 26·0 | 20·3 | 51·4 | 932 |
| 1811–20 | 13·1 | 9·7 | 29·2 | 22·2 | 52·0 | 1,100 |
| 1821–30 | 16·3 | 8·3 | 31·9 | 24·8 | 56·4 | 1,485 |
| 1831–40 | 15·2 | 11·4 | 33·3 | 20·5 | 60·0 | 2,466 |
| 1841–50 | 16·8 | 8·9 | 23·9 | 16·8 | 49·6 | 4,223 |

*Notes*

a  'Total No. of patents sealed' is a net total which excludes patents registered in the name of patent agents. 'Communications' from third parties registered in the name of multiple patentees are also excluded.

b  Double counting has been avoided in the case of joint and triple patents, etc., by weighting the patents accordingly. In the case, for example, of a triple patent, each inventor is credited only with a third of the patent.

c  Cross figures, excluding patents registered in the name of patent agents, including communications registered in the names of multiple patentees and dispensing with a weighting system, yield a marginally higher percentage of total patents attributable to multiple patentees.

*Sources.* B. Woodcroft, *Alphabetical Index of Patentees of Inventions*, 1969 edition, and *Chronological Index of Patents of Invention*, 1854.

assets with variable yields and risk. If inventors are attempting to maximise profits within a risk constraint, then in a particularly high-risk occupation such as inventing they should attempt to reduce risk by spreading their inventive effort. This might be done by inventing in diverse areas in one industry, for example, spinning and bleaching in the cotton industry, or by inventing across several industries. Diversification, moreover, also implies that inventors expected profitability to vary between industries over time, and that they allocated their resources accordingly.[59]

The extent to which inventors diversified their inventive portfolio depends on how an industry is defined. This is a major problem, which is not easy to handle. A narrow definition, such as that employed in the Standard Industrial Classification, poses one serious difficulty: can it really be argued, for example, that the inventor who patented a device for making pins, screws, rivets and nails was inventing in four quite separate industries? The broader classification used by Bennet

Woodcroft for the Patent Abridgement Specifications, which groups patents enrolled prior to 1852 under 103 industrial headings by process, is only slightly more helpful and is spoiled by the fact that a considerable number of patents appear under two or more headings.[60] Many inventions in the cotton industry, for example, were double-counted because they applied to both spinning and weaving, two quite separate categories in Woodcroft's abridgements. Given these problems, a much broader definition of industry than that conventionally employed has been adopted for present purposes. Thus all textile-related patents have been grouped under one head, marine patents under another, motive power under a third, etc. This broad system of classifying industries has been adopted not because it is intrinsically better than, say, Woodcroft's but because it necessarily yields a conservative estimate of diversification. This should disarm critics who might otherwise argue that the evidence tends to exaggerate the extent to which inventors were prepared to diversify.

The data relating to diversification are contained in tables 6 and 7.

Table 6  *Industrial spread of multiple patentees, 1751–1852*

| Patents per patentee | No. of patentees | No. of industries |       |      |      |      |      |
|---|---|---|---|---|---|---|---|
|   |   | 1 | 2 | 3 | 4 | 5 | 6+ |
| 2 | 1,180 | 674 | 506 |   |   |   |   |
| 3 | 440 | 170 | 176 | 94 |   |   |   |
| 4 | 223 | 62 | 68 | 54 | 39 |   |   |
| 5 | 114 | 28 | 26 | 29 | 20 | 11 |   |
| 6 | 85 | 19 | 16 | 17 | 17 | 12 | 4 |
| 7 | 44 | 5 | 11 | 5 | 14 | 3 | 6 |
| 8 | 30 | 4 | 8 | 5 | 6 | 3 | 4 |
| 9 | 16 | 2 | 1 | 3 | 2 | 2 | 6 |
| 10 | 15 | 2 | 2 | 3 | 1 | 4 | 3 |
| 11+ | 54 | 4 | 3 | 3 | 9 | 7 | 28 |
| Total | 2,201 | 970 | 817 | 213 | 108 | 42 | 51 |
| % | 100 | 44·1 | 37·1 | 9·7 | 4·9 | 1·9 | 2·3 |

Note. Exclusions as note (a), table 5.
Source. Woodcroft, *op. cit.*

*Table 7   Industrial spread of multiple patentees holding four or more patents*

| No. of patentees Holding 4+ | As % of all multiples | \multicolumn{6}{c}{No. of industries} |
|---|---|---|---|---|---|---|---|
| | | 1 | 2 | 3 | 4 | 5 | 6 |
| 581 | 26·4 | 126 | 135 | 119 | 108 | 42 | 51 |
| % of all 4+ | | 21·7 | 23·2 | 20·5 | 18·6 | 7·2 | 8·8 |

Source. Woodcroft, *op. cit.*

The most notable feature of table 6 is that only 44·4% of all multiple patentees active in the period 1751–1852 confined their attention to one industry.[61] Conversely, 40·5% of all patentees holding three or more patents invented in at least three industries. The extent to which the quasi-professional inventor diversified his inventive portfolio may be more clearly seen in table 7, where, of the 581 patentees taking out four or more patents, only 21·7% confined their attention to one industry. Relatively few of this group of patentees, it is true, invented in more than four industries: but, as Sharpe has pointed out, a little diversification goes a long way.[62] The reduction in risk resulting from the diversification that did take place must, therefore, have been substantial.

The preferred areas of invention for multiple patentees, as for patentees as a whole, were textiles and motive power. Although as the industrial revolution progressed some inventors were to become constrained in their choice of area by the extent of their skills and knowledge, many were able to switch their resources by diversifying *within* a particular industry. Textiles are a case in point. During the eighteenth century 44% of textile patents were taken out for bleaching, dyeing, printing and finishing. Patents for production processes increased only from the mid-century and, apart from the 1750s and 1760s, spinning was always more important. From the mid-1790s spinning remained the dominant area for patent activity in textiles, with a sharply rising trend from 1765 to the turn of the century, after which it begins to flatten. Weaving, on the other hand, apart from two peaks in 1760 and 1805, did not experience a distinct increase in patent activity until the spread of the improved power loom in the

1830s and 1840s. Patenting in finishing continues to be important through to the 1830s, after which it declines relatively. Overall this intra-industry diversification confirms that inventors were able to switch their resources as the nature of production changed. In the eighteenth century when textiles were organised on a putting-out system the growth in output depended upon changes in the nature of products, rather more than changes in the means of producing those products: markets dominated production. With the emergence of the factory system patent activity in production processes became more significant, although patents for designs, patterns and colours still remained important.[63] 'Heroic' inventors were equally adept at diversifying, and whilst they tended to exhibit a slightly greater predilection for textiles and motive power than patentees as a whole, they spread their activities over a wide range of industries.[64] This again reflects the rudimentary nature of much technology in the eighteenth and nineteenth centuries, an age when the skills possessed by a barber, a carpenter and a parson were sufficient to transform a national industry.[65]

## Notes

1  J. James, *History of the Worsted Manufacture in England*, 1857, p. 333.
2  A. Toynbee, *Lectures on the Industrial Revolution in England*, 1969 ed., p. 90.
3  J. H. Clapham, *An Economic History of Modern Britain: the Early Railway Age*, Cambridge, 1939, p. 143.
4  Increases in aggregate inputs, allegedly, explain only a small part of the measured growth of output. Consequently, the main source of growth has to be explained by 'residual' factors. Technology is claimed to be one of the most important components of the residual.
5  S. Lilley, *Technological Progress and the Industrial Revolution, 1700–1914*, 1970, p. 5. This terminology comes from Marx.
6  Landes, *op. cit.*, p. 41.
7  W. W. Rostow, 'The beginnings of modern economic growth in Europe: an essay in synthesis', *Journal of Economic History*, 1973, p. 570.
8  For an exception see Musson and Robinson, *op. cit.*, 1969. It is important to note, though, that Musson and Robinson are more concerned with the relationship between science and industry than with invention *per se*.
9  See, for example, G. N. von Tunzelmann, *Steam Power and British Industrialization to 1860*, Oxford, 1978. 'Like most economic historians, I have paid far more attention to the imitation and the diffusion of a technique than to its original invention', p. 295; and C. K. Hyde, *Technological Change and the British Iron Industry, 1700–1870*, Princeton, 1977. 'I will devote little time to the *invention* of new techniques but will concentrate on their diffusion', p. 3.

## Patents and inventive activity

10 C. P. Kindleberger, 'The historical background: Adam Smith and the industrial revolution', in Wilson and Skinner, *op. cit.*, p. 16.
11 K. Norris and J. Vaizey, *The Economics of Research and Technology*, 1973, p. 40; E. Mansfield, *The Economics of Technological Change*, 1968, p. 51; C. F. Carter and B. R. Williams, *Investment in Innovation*, 1958, p. 15; W. Bowden, *Industrial Society in England towards the end of the Eighteenth Century*, New York, 1925, p. 29; T. S. Ashton, *The Industrial Revolution*, Oxford, 1962 ed., pp. 11–13; P. Mathias, *The First Industrial Nation*, 1969, p. 136; P. Deane, *The First Industrial Revolution*, Cambridge, 1969, p. 95. What follows is not meant to be a comprehensive survey. For this, see C. Kennedy and A. P. Thirlwall, 'Surveys in applied economics: technology', *Economic Journal*, 1972, pp. 11–72; Introduction by A. E. Musson, ed., *Science, Technology and Economic Growth in the Eighteenth Century*, 1972, pp. 1–68; R. R. Nelson, 'A survey of the economics of invention', *Journal of Business*, 1959, pp. 101–27.
12 F. Taussig, *Inventors and Money-makers*, New York, 1915; Jewkes *et al.*, *op. cit.*; J. Rossman, *The Psychology of the Inventor*, Washington, D.C., 1931; F. Vaughan, *Economics of the Patent System*, New York, 1925, pp. 1–10.
13 See, for example, N. F. R. Crafts, 'Industrial revolution in England and France: some thoughts on the question "Why was England first?"', *Economic History Review*, 1977, p. 434.
14 One notable exception is F. Neymeyer, *The Employed Inventor in the United States*, Massachusetts, 1971. This examines the relationship in a modern context.
15 J. Schumpeter, *Business Cycles*, New York, 1939, I, p. 84.
16 A. Pigou, *Economics of Welfare*, 1924 ed., p. 163; see also *Industrial Fluctuations*, 1927, pp. 41–4; Sir J. Stamp, *Some Economic Factors in Modern Life*, 1929, p. 113; see also *Invention as an Economic Factor*, the Watt Anniversary Lecture, 1928, pp. 25–6.
17 S. C. Gilfillan, *Sociology of Invention*, Chicago, 1935; see also 'Invention as a factor in economic history', *Journal of Economic History*, 1945, pp. 66–85; W. F. Ogburn, *Social Change*, New York, 1922; W. B. Kaempffret, *Inventions and Society*, Chicago, 1930.
18 See also Z. Griliches and L. Hurwicz, eds., *Patents, Invention and Economic Change*, Harvard, 1972. This includes a bibliography of J. Schmookler's writings.
19 Jewkes *et al.*, *op. cit.*, pp. 210–11.
20 N. Rosenberg, 'Science invention and economic growth', *Economic Journal*, 1974, p. 98. This is also the central thesis of Musson and Robinson, *op. cit.*, 1969, where they considerably modify 'the traditional view of the Industrial Revolution as being almost entirely a product of uneducated empiricism', p. vii.
21 P. S. Johnson, *The Economics of Invention and Innovation: with a case study of the development of the Hovercraft*, 1975, p. 37.
22 Schmookler, *op. cit.*, pp. 108–9.
23 It is important to note that Schmookler's analysis does not depend on inventors acting as profit maximisers. He simply assumes that 'the

probability that any given possible invention will be made varies directly with its expected profitability ... This does not imply that inventors are income maximisers. It implies only that the higher the expected returns to inventing, the more likely they are to invent than do something else; and that they are more likely to make invention *a* than invention *b*, the higher the expected returns from *a* relative to those from *b*,' ibid., p. 114.

24 A. Smith, *An Inquiry into the Nature and Causes of the Wealth of Nations*, 1976 ed., I, p. 342.
25 *Ibid.*, I, p. 128.
26 Dircks, *op. cit.*, 1867, p. 80.
27 *Mechanics Magazine*, XXV, 1836, p. 456.
28 Musson, *op. cit.*, 1972, pp. 52–3.
29 Rosenberg, *op. cit.*, 1976, pp. 62–6.
30 N. Rosenberg, 'On technological expectations', *Economic Journal*, 1976, pp. 523–35.
31 Musson and Robinson, *op. cit.*, 1969, p. 251.
32 Babbage, *op. cit.*, 1830, pp. 132–3.
33 C. S. Drewry, *Observations on the Defects of the Law of Patents*, 1863, p. 7.
34 H. Bessemer, *Autobiography*, 1905, p. 82.
35 Musson and Robinson, *op. cit.*, 1969, p. 207.
36 Boulton and Watt MSS, Watt to Lord Loughborough, 8.7.1785, Letter Book, B.R.L. This letter was written shortly after Arkwright's patent had been set aside, and whilst the Irish proposition was being discussed.
37 See for example, F. M. Scherer, 'Patent statistics as a measure of technical change', *Journal of Political Economy*, 1969, pp. 392–8; S. Encel and A. Inglis, 'Patents, invention and economic progress', *Economic Record*, 1966, pp. 572–88; J. Schmookler, 'Interpretation of patent statistics', *Journal of the Patent Office Society*, 1950, pp. 123–46, and 'The utility of patent statistics, *Journal of the Patent Office Society*, 1953, pp. 407–12.
38 *Op. cit.*, 1967, pp. 407–12.
39 *Op. cit.*, p. 188.
40 *Select Committee on Patents*, Parl. Papers, XVIII, 1851, p. 329.
41 See above. This was less than inventors paid in total after the 1852 Act, although then they were able to pay by instalments. Badwell MSS, T. Gill to J. Brackenbury, 9.7.1852 (MSS4), Salford University Library; *Select Committee on Patents*, Parl. Papers, XVIII, 1851, p. 429.
42 Kendrick, *op. cit.*, 1774, p. 38.
43 *Hansard*, XXI, 1824, pp. 598–608; Brougham MSS, [22,371], D. Booth to Lord Brougham, 22.10.1841, U.C.L.; for a contrary view see Musson and Robinson, *op. cit.*, 1969, especially chs. II, X.
44 Cf. Edwards, *op. cit.*, p. 56.
45 B. Woodcroft, *Brief Biographies of Inventors of Machines for the Manufacture of Textile Fabrics*, 1863, p. 56.
46 Brougham MSS, [3,085], E. F. Herrington to Lord Brougham, 27.5.1848, U.C.L.; *Select Committee on Patents*, Parl. Papers, XVIII, 1851, p. 287; Kindleberger, *op. cit.*, p. 22. Industrial espionage during the industrial revolution is a subject which warrants an extensive study.

47 R. C. Post, '"Liberalizers" versus "scientific men" in the antebellum Patent Office', *Technology and Culture*, 1976, p. 54.
48 In fact, it is quite possible that the number of patents actually overstates the volume of inventive activity, if patents are taken out for inventions which have been copied. On the other hand, some industries where ideas were not embodied in machinery were less likely to use patent protection. See, for example, W. W. Bladen, 'Potteries in the industrial revolution', *Economic History*, 1926, pp. 117–30.
49 *Observations on the Utility of Patents*, 1791, p. 53.
50 *Hansard*, X, 1824, p. 145.
51 *Transactions of the Society of Arts*, LV, 1843–44, p. XV.
52 *Select Committee on Patents*, Parl. Papers, XVIII, 1851, pp. 232, 398.
53 *Mechanics Magazine*, V, 1859, p. 69; see also Brougham MSS, [3,085], E. F. Harrington to Brougham, 27.5.1848, U.C.L.
54 *The Patent: a Poem*, by the Author of the Graces, 1776.
55 *Patent Journal*, I, 1846, p. 9.
56 It is very difficult to think of 1,000 inventions which were not patented. The sample of 13,277 patents is large enough to support the foregoing analysis. It also seems likely that the proportion of inventions patented increased from the 1780s as competition in the invention industry intensified, and as inventors learned the value of having a patent. For unpatented inventions between 1660 and 1750 see MacLeod, *op. cit.*, pp. 165–205.
57 These calculations are based on B. Woodcroft's *Alphabetical Index of Patentees of Invention*, 1854 and his *Chronological Index of Patents of Invention*, 1854, both of which terminate in 1852. The multiple patentees who took out only one patent prior to 1852 would necessarily appear as a single patentee.
58 Obviously, it would be interesting to know exactly what proportion of important inventions were produced by multiple inventors, but such an analysis raises a number of problems. Although almost everyone has some intuitive sense of what an important invention is, how is it to be judged so? 'Should it be judged according to its superiority over earlier best practice or according to its usually much smaller margin over other inventions made synchronously with it?' Or should it be judged by the extent of its commercial exploitation? None of these questions is easily answered. In one sense, however, these problems are less important than they first seem. Even though there is very little doubt that some inventions were more important than others (and some were undoubtedly trivial), this study is concerned with inventive activity generally and therefore should take into account all inventions, minor ones as well as major ones. Both involved the use of scarce resources and in most cases genuinely required inventive effort. Schmookler, *op. cit.*, p. 19.
59 Patent activity was fairly evenly spread between industries during the industrial revolution, Appendix A. For a more extensive analysis see H. I. Dutton, *The Patent System and Inventive Activity during the Industrial Revolution, 1750–1852*, unpublished Ph.D., University of London, 1981.

# Invention and inventive activity 121

60 Classifying inventions by process will automatically lead to double counting because they can be used in a number of industries.
61 This figure is certainly inflated by the fact that the post-1852 activities of multiple patentees active prior to 1852 have been ignored.
62 W. F. Sharpe, *Portfolio Theory and Capital Markets*, New York, 1970, p. 130. I would like to thank Dr Keith Jones, University of Auckland, for this reference.
63 D. C. Coleman, 'Textile growth', in N. B. Harte and K. G. Ponting, eds., *Textile History and Economic History: Essays in Honour of Miss Julia de Lacy Mann*, Manchester, 1973, pp. 1–21.
64 I.e. inventors holding at least eight patents in four or more industries.
65 Coleman, *op. cit.*, 1973, p. 13.

# 7

# Trade in invention

The existence of an active trade in invention is the final criterion that has to be satisfied. To show that such a trade existed, it is necessary to demonstrate that inventors were sellers as well as users of inventive output, and that this output was usually bought and sold in the market place. Another, but less necessary, condition implied by the trade in invention is that inventive activity increasingly became a specialised trade, in the sense that the production of inventive output depended more and more on those who worked *outside* the industries which use inventive output. Unfortunately, the evidence is rather patchy and it is difficult to know whether it is typical. This chapter will, therefore, provide only a number of illustrations.

That there was a market for invention and that inventors were frequently sellers, rather than users, of inventive output is confirmed by a number of informed contemporaries. This point was well made by Adam Smith, despite his alleged ignorance of the recent technological advances made in cotton, iron and steam:[1]

> All improvements in machinery, however, have by no means been the invention of those who had occasion to use the machines. Many improvements have been made by the ingenuity of makers of machines, when to make them became the business of a peculiar trade: and some by that of those who are called philosophers, or men of speculation, whose trade it is, not to do anything, but to observe everything, and who upon that account, are often capable of combining together the powers of the most distant and dissimilar objects. In the progress of society, philosophy or speculation becomes, like every other employment, the principal or sole trade and occupation of a particular class or citizen.[2]

The view that a great many inventions, especially important ones, came from outside the user industry has been so frequently repeated it cannot be ignored. F. J. Bramwell, president of the Institute of Mechanical Engineers, wrote: 'The fact is, that the bulk, one might say the whole, of real substantive inventions have been made by persons not engaged in the particular pursuit to which the inventions relate.'[3] 'It is a remarkable fact,' Henry Dircks observed, 'that inventions more frequently than otherwise come from men disconnected with the trade to which they appeal for patronage.'[4] William Carpmael, who knew more about the invention market than most, argued exactly the same: 'it will be found, that in most instances, where an extensive change has been introduced, it has been by persons before unconnected with the particular branch of manufacture'.[5] Bessemer, whose enthusiastic amateurism fast gave way 'to a more steady commercial instinct',[6] argued the case with force: 'I find that persons wholly unconnected with any particular trade ... are men who eventually produce the great changes.'[7] The extent to which inventors diversified their invention portfolios across several industries also supports the importance of what has been called the 'uncommitted mind'.

In the cotton and woollen industries there is some evidence to support these judgements. Between 1790 and 1830 virtually 50% of textile inventions were patented by those who were not manufacturers or artisans, and although users were relatively more important as producers of inventive output especially between 1825 and 1830, merchants and gentlemen still maintained an active interest in inventing, as did machine makers in the capital goods industry. Patentees from totally unrelated trades — brewing, teaching and the Church — were the only group to experience a precipitous decline, and by 1825 they had disappeared as technology became more complex and costly.[8]

The notion that inventing became a specialised trade or an industry in its own right is a much more contentious view. It is almost certain that few, if any, inventors did nothing else but invent. It is equally certain that manufacturers and users of inventions succeeded in producing inventive output to counteract rising costs, competition and specific technical problems, or all three. In between, though, there appears to have been a substantial degree of specialisation, and there are two reasons why this should have been so.

Firstly, inventors were able to free themselves from the day-to-day problems which confronted the ordinary businessman. Inventors 'as a class' were, as many observed, 'singularly deficient in the qualifications for prosecuting a new trade'.[9] Consequently, inventors were generally reluctant to dissipate their efforts in the commercial exploitation of their inventive output, because the fates of insolvency were too easily tempted. Secondly, manufacturers and users of inventions were more than satisfied with the arrangement. It allowed them to concentrate on their own business affairs, and enabled them to minimise the resources allocated to the production of invention, an activity where men without talent or skill were not likely to reap high returns. Apart from making minor improvements, many users were unwilling to indulge in extensive inventive activity. 'Trying to invent is a great mistake', one Dundee textile manufacturer observed, 'Our vocations as Captains of industry is to work our business so as to produce the best article at the cheapest rate ... if in the prosecution of this object we fall upon something new which we can turn to account it is fair gain. But to lay yourself out for inventing something new is a mockery, a delusion and a snare into which many wise men have fallen.'[10] Even where users were able to identify the need for technological change many, it seems, preferred to buy rather than to produce their own inventions. Few, if any, were prepared to buy inventions without the protection of a patent.[11]

This specialisation which was a general characteristic of the economy during the early nineteenth century was obviously far from complete, and few inventors were what Brunel called 'professional inventors'[12] or what Smith called 'philosophers'. Many workmen who invented remained workmen resisting the temptation to 'professionalise' their infrequent dabblings, although even in these circumstances they still attempted to sell their output.[13] Similarly, there were manufacturers who produced their own inventions. To argue that all inventive output was produced by inventors who worked exclusively in an invention industry would be unjustified. Yet the existence of an ever growing class of inventors, specialising in the production of inventive output for other users, cannot be ignored. Contemporaries were quick to sense what was happening. Inventing and the introduction of patented inventions, Thomas Webster stated, had become 'in fact a sort of trade or business. In consequence of the knowledge that protection will afford, scientific people are induced to take to invention as a pursuit and means of livelihood ... from the knowledge that they can make it a business and obtain a profit by doing so.'[14] The developing invention industry was regarded by contemporaries in virtually the same way as other industries.

The market for inventive output was quite complex, and inventors could dispose of their inventions in a number of ways.[15] They could sell the whole or part of their inventions; they could license their inventions to one or more users; they could form partnerships with users and grant or refuse to grant licences to other users; finally, they could arrange with one or more individuals to trade for the benefit of all parties involved.

It is difficult to say which was the most common. The method varied between inventors and depended upon the structure of the user industry. Selling the rights of inventions was probably the method which quasi-professional inventors preferred, since it allowed them to concentrate on inventing. An exclusive sale was also the most likely way to persuade users of an invention's value.[16] Old customs and prejudices frequently made manufacturers resist new ideas. Unless they were promised sole-user rights, they would continue with well tried techniques.

To estimate the extent to which patents were sold would demand an exhaustive search of every patented invention, which obviously cannot be done here. There are, however, two crude measures, and although neither is in itself adequate, they give some indication of the scale of selling inventive output. The first measure is derived from the number of patents contested in the courts. Out of 203 patents contested, approximately 19% were assigned by inventors to users. Significantly, the proportion of patents sold changed over the period of the industrial revolution. Between 1770 and 1799 only 7% of contested patents were sold; between 1800 and 1829 this had increased to 18%, and in 1830–39 and 1840–49 the proportion was 22%. Since, as will be seen later, only important inventions were likely to come before the courts, this sample of patents may overstate the total number sold.[17] The second measure, which is derived from the number of petitions examined by the Judicial Committee of the Privy Council between 1835 and 1852 to prolong the duration of patents, yields a rather similar proportion. Of the seventy-five petitions examined, 25% had been assigned by inventors to users.[18]

The claim that 'many patents for important inventions are assigned'[19] is supported by other evidence. Few inventors, though, were driven to sell their patents by auction, as William Collins threatened to do in 1808 after several Birmingham manufacturers had delayed producing his ventilators.[20] Most would approach firms to see what price they would be prepared to pay. Bessemer, for example, sold a number of his patents in this way. When he attempted to sell his machine for cutting lead pencils to the leading pencil makers, Mardon and Co., he notes in his autobiography, 'I fear this little episode does not speak very favourably for my business capacity, for I certainly ought to have made much more than I did by this really important invention.'[21] His invention for bronze powder was rather more successful,[22] but in the 1840s he was unable to interest glass makers in an improved method of manufacturing plate and flint glass which he patented jointly with Schonberg.[23] In the flax industry Marshall's were, after some anxious moments, quite happy to buy for £20,000 the exclusive rights to the combing machine invented by Heilman, because they thought it would 'renew our good name' and rejuvenate the firm as the wet spinning invention had done in the early 1830s: 'everything turns on the real value of this invention, if it is what it professes, I do not see how we can do otherwise'.[24] In 1835 David Mushet assigned his 1835 patent to William Mushet for £15,000,[25] and some years later Hardy's patent for railway axles was sold for £3,000 and, within a period of four years, the buyers realised a sum of £23,000.[26] It would be tedious to continue listing inventors who attempted to sell the rights to their output and, although a case study approach may not be representative, inventive activity in the pin industry offers a striking and interesting illustration. Here all inventions made between 1800 and 1835, with one exception, were produced by men from outside the industry, and all attempted to sell their output to pin manufacturers.

The small domestic pin workshops which Smith wrote about had almost disappeared by the mid-eighteenth century.[27] At the turn of the century, pin manufacturing was largely confined to Gloucester, Bristol, Warrington and Birmingham. Between 1800 and 1840 the number of firms fell from roughly twenty to around half a dozen. By 1860 the trade was dominated by three producers, Edelston and Williams of Birmingham being the largest. Over this period, new techniques slowly transformed pin making from a hand-tool labour-intensive activity into one using automatic and complex machines.

Inventive activity concentrated on two stages of pin making, heading and pointing, which together made up 57% of total cost. Heading was the most expensive and most labour-intensive process.[28] It was done by women and children in the home (especially during periods of excess demand), and in the factory.[29] Heading required the operative to spin a coil of wire round a shank and to cut every three or four turns. This segment — a spun head — was then slipped over a pin shaft, placed horizontally on to a die, and fastened to it with repeated blows of a hammer, whilst turning the pin between finger and thumb. It demanded nimbleness and dexterity. Pointing, a highly paid male activity, was in contrast a process which required skill and strength, and took place almost entirely within the factory. Between

1780 and 1852 eighteen patents were taken out to economise on the use of labour in both processes.[30]

Down to 1852 two types of invention were produced. The first sought to save labour and improve the quality of pins, without changing the product in a significant way. The second also saved labour by the introduction of automatic machinery, but changed the product, from a pin in which the head and shank were made from separate pieces of wire to a pin with a solid head, that is, a pin in which the head and shank were formed from the same piece of wire. This new product solved the problem of heads becoming detached from the shank.

The first three inventions patented after the *Wealth of Nations* can be explained largely by labour shortages and rising wages during the Napoleonic wars. In 1797 Timothy Harris patented a method of casting a lead-based alloy head on to an iron shank,[31] but this does not appear to have been taken up by the trade. Unfortunately little is known about the invention, or whether Harris attempted to sell it to pin makers. Harris, who also took out an omnibus patent in 1814 for processes in the paper, silk, linen, woollen, cotton and leather industries, was certainly not a pin inventor nor, it seems, a prolific inventor.[32]

More is known of William Bundy, who invented the 'Patent' pin machine in 1809. Bundy was a mathematical instrument maker and an inventor of some note. Between 1796 and 1830 he took out ten patents, ranging from machinery designed to cut and make combs to an anti-evaporation cooler to regulate the temperature of warts and wash in fermentation.[33] His pin machine, which was an extension of the foot stamping apparatus — an allegedly barbarous contrivance — riveted heads on to the pin in a single blow and was therefore much easier and quicker to operate. Heads and shanks were still prepared separately, and the machine was not widely adopted until after the patent lapsed in 1823. Durnford's, one of the more progressive firms, appear to have adopted the machine before 1818, and Hall, English and Co., another Gloucester manufacturer, installed it in 1824.[34] Whether Bundy restricted the use of the machine or set a prohibitively high royalty is unknown, as are his dealings with other pin manufacturers. But since his machine frequently broke down, and was only suitable for installation in a factory, pin manufacturers used to the flexibility of sub-contracting out work to adjust capacity may have been reluctant to switch completely to the new technology.

The patent pin machine was no more than a sophisticated tool. The invention patented in 1812 by John Leigh Bradbury, an engraver and calico printer, and Charles Weaver, a pin maker from Gloucester, was an altogether more complicated affair which combined several moving parts to attach a ready-spun head to a pointed shank.[35] It is very difficult to see who was the main source of this idea, but Bradbury patented five other inventions between 1807 and 1824, three for spinning fibres and two for engraving.[36] He also developed a machine which printed banknotes, but for obvious reasons this was not patented. Whether Bradbury formed a partnership with Weaver to share risk, to acquire capital or to gain both inventive and innovational profits is unclear, but in 1816, after failing to interest pin makers, he set up as a calico printer.[37] In the next year the unfortunate Weaver — the only pin manufacturer involved in inventing — went bankrupt.

In 1817 Seth Hunt, an American living in London, took out a patent for a machine which provided almost all the basic features of subsequent pin inventions.[38] Although his machine was capable of pointing, it was essentially designed to produce a solid headed pin. Hunt soon attempted to sell the invention. Three months before his patent was sealed (it is likely that it was passing through the Patent Office) he approached the Warrington firm of Stubs Wood and offered them the sole rights of the patent for a three-year period, at a cost of £7,500.[39] At a time of falling demand and financial stringency the cost proved prohibitive, although Hunt ultimately induced Cowcher, Kirby and Co., the largest pin-making firm, to take up the machine. With the exception of a brief flurry in 1818 the pin trade remained depressed until 1821 and Hunt, who took out one more patent in 1817, for an escapement mechanism, disappears from the record. Cowcher, Kirby and Co., continued to perfect the machine, but it was not commercially viable until the late 1830s. Cowcher's were at first obliged to employ Lemuel Wright, an American engineer living in England, to make the machine work, but he soon left the firm to develop his own pin-making machine.[40]

Above all others involved in the development of pin machinery Wright was *the* quasi-professional inventor, taking out no fewer than twenty patents over a period of twenty-nine years. Starting with machinery for making bricks and tiles, patented in 1820, he moved to pins, then to steam engines, patents for bleaching fabrics, and a chimney sweeping device, until in 1848 he took out his final patent, a process for preparing fibres for his spinning machine.[41] Wright took his inventing seriously, and by 1826 he already had an 'immense manufactory, two floors of which ... are intended for manufacturing steam engines and various sorts of machinery ... They say that £30,000 has been expended already.'[42]

In 1822, two years before he registered his patent, Wright attempted to sell his ideas to Hall, English and Co. He wanted £900 for six machines and a royalty of £30 per annum. Hall found the terms unacceptable, especially as Wright would not let him see a machine at work, had not offered him exclusive use, and had no patent to provide protection.[43] Wright had no more success the following year when he approached Stubs Wood.[44]

Nothing more is heard of Wright's activities in pins until 1826 — two years after his patent was sealed — when he was in the process of negotiating with Cowcher, Kirby and Co.[45] Most other manufacturers seem to have believed that Wright's machine 'would never pay'.[46] The durability of capital in the shape of the patent pin machine, the relative abundance of labour and the stringent financial conditions made innovation risky and expensive. These conditions prevailed until the late 1820s. In the interim Wright assigned his machine for bleaching and washing to the Manchester firm Bower and Bark, took out another six patents, and in 1829 was declared bankrupt.

In 1829 Wright did, however, succeed in selling his pin patent for £3,775 to D. F. Tayler, later of the Stroud firm, Tayler, Shuttleworth and Watnerby, mechanic, capitalist and book-keeper respectively. Wright, though, was contractually committed to the Stroud firm to make the machine work and was unavoidably diverted from inventing. Between September 1828 and December 1833 there are no patents registered in his name.[47] By late 1832 the new solid headed pins were ready to enter the trade, and within two years were successfully 'playing devil' in the London market.[48]

With the revival of the pin trade in the early 1830s several other inventors approached the main manufacturers. In autumn 1831 M. Amédée Raymond, a French inventor residing in Birmingham, wanted £300 and £25 for each of his solid heading machines, but neither Hall and English nor Durnford's showed much interest.[49] William Hunt, one of Wright's ex-mechanics, had no more success, despite some initially favourable comments and a growing desire to acquire machinery which would enable manufacturers to produce a competitively priced solid headed pin.[50] Another machine was invented by Daniel Ledsam and William Jones, both from Birmingham, but neither of them was a pin maker. Ledsam had previously invented a nail machine, whilst Jones had three other patents for the manufacture of iron, nails and filtering fluids.[51] Despite approval of their sample pins, no firm took the invention up. Violent fluctuations in American demand,[52] the major export market, made firms extremely cautious, especially since there were still some technical difficulties to overcome.

From the mid-1830s all pin inventions were made by Americans and mostly in America. The two main inventors were again not originally pin manufacturers. Samuel Slocum, who allegedly invented his solid-head machine in England[53] and attempted to sell it there between 1831 and 1835, came from Rhode Island, and was trained as a carpenter. He invented and patented a nail-making machine in 1834 and 1835[54] but, finding no market for any of his inventive output, returned to the United States, where, in partnership with Gellson, he opened a pin manufactory at Poughkeepsie, New York. By the early 1840s his pins were causing American pin importers some concern.[55]

John I. Howe, who patented a pin machine in America in 1832, had a similar experience. Howe was a physician by profession, whose knowledge of chemistry had led him to experiment with a rubber compound, and in 1828 he patented a process for making India-rubber. He also constructed a machine to process his rubber mixture, and was attracted to inventing pin machines only after his rubber venture failed in the winter of 1830. Howe, who as medical officer at the Belleville almshouse in New York regularly observed the inmates making pins, first invented a machine which produced spun (not solid) heads. He later sold a share of his patent to the New York merchants Jarvis Brush and Edward Cook, who in return reimbursed him for his past expenses and agreed to finance all further developments. After failing to sell his machine whilst visiting London and Manchester in 1833 and 1834, he returned to the United States and established the Howe Manufacturing Co. in 1835. Three years later he successfully converted his machine to solid head production, and in 1840 patented the Rotary Pin Machine.[56]

One significant feature of this case is that the search process was carried out almost exclusively by inventors hawking their inventions from firm to firm in an attempt to create a demand for their output. Pin inventions did not simply respond to a *specific* demand, and nearly all the inventors devoted a great deal of time and energy trying to break down the manufacturers' commitment to routinised production.[57] Pin producers themselves hardly ever attempted to invent, and generally waited to be approached by the sellers of inventive output.

In some other industries firms were more active in searching for inventions, although very few 'institutionalised' the search activity. Price's Candle Co., which had the biggest candle factory in the world, was the largest holder of patents for treating sterine.[58] W. S. Hales estimated that 'with what they have purchased and what they have taken out, they possess now from eighteen to twenty patents. Separately they would be of little use but collectively it is certainly an important improvement.'[59] In the 1840s the company wanted to purchase A. V. Newton's (the patent agent) 1841 patent for distilling palm oil, but he would not accept the £1,000 offered and Price and Co. threatened to cancel the patent by a writ of *Scire facias*.[60] Hale was himself in the process of negotiating a 'treaty for one or two [patents] which in themselves are useless, yet they contain the germ of something, and it is worthwhile, if I can get them for a small sum, but directly you make application for a patent ... it becomes very valuable all at once'.[61] The Electrical Telegraph Co., which bought Wheatstone's patent for £30,000, was another firm which actively attempted to buy all the patents which could interfere with its competitive position.[62] J. L. Ricardo claimed that the firm made it a rule to buy any closely related invention if offered reasonable terms, because it was much cheaper than bringing an action against anyone who set up against them.[63] In the rubber industry Charles Hancock and George Spencer, the two leading producers, bought patents for much the same reason, and to maintain their grip on the market.[64] It is also interesting to note that Spencer-Moulton's frequently refused to take up licences offered by inventors unconnected with railways but never declined 'an interest in inventions patented by men whose influence was important in that sphere'.[65] Chance, the glass maker, was another user who approached an inventor, in this case Bessemer, to buy his reverberatory and cylindrical rotating crucible for making sheet glass. He was willing to pay £6,000 as long as there was no discussion over the price: 'Just think it over and determine whether you will sell me your invention as it stands and make at once a profit on what you have done, or whether you will spend more labour and money with the chance of much greater remuneration if you succeed commercially and no one else supersedes you.'[66]

How many firms were active in this way remains unknown, but they seem to have been the exception. When Richard Roberts was asked if he knew of firms who frequently bought up patents to forestall competition, potential and actual, he could only reply that he had heard of that kind of thing but had never suffered from it.[67] Significantly,

Webster refers only to candles and rubber as industries which were 'created and built up in short time' because of patents.[68] For most industries, especially old ones with a large number of producers, buying up a large number of patents was impossible, and although many users were prepared to watch closely what was happening in the invention industry, few actively sought out sellers. From the evidence it seems that inventors were generally the active party and, as in any market, some were more successful than others.

Another way inventors could trade their inventions was through a system of licensing.[69] Richard Roberts thought this the 'best and most commercial' method of disposing of inventions.[70] It was perhaps also the most complex. Inventors needed to have a good sense of what the market would bear. They had to calculate, as far as it was possible, the optimum royalty rate and the optimum number of firms to license, none of which was easy to do, despite the profusion of advice offered by patent agents and other writers on the subject. The types of licensing agreement, consequently, varied quite considerably from invention to invention and from industry to industry, and often depended upon the personalities of the parties striking the bargain.

Although the grant of a licence meant that the patentee retained ownership, some licence agreements were equivalent to the direct sale of a patent to a single user. In 1845 Elias Robinson Handcock, agreed to grant a sole licence to Samuel Lloyd and partners, iron manufacturers at Old Park Iron Works, Wednesbury, for the manufacture of turntables for railways.[71] Rupert Ingleby, a copper smelter, was likewise the only licensee of Hollingrake's patent,[72] and Marlowe, Allicott and Seyrig of Nottingham held the sole licence for the manufacture of centrifugal machines used in sugar refining.[73] How many inventors were prepared to offer sole licences is unknown, but many users were unwilling to take out a licence unless they were given exclusive use. Roger Hopkins of the Monmouthshire Iron and Coal Co., in lower Ebbw Vale, was not prepared to pay David Mushet 1*s* per ton for his patent for making bar iron 'unless we have an exclusive right in your patent for Wales. I should not feel myself justified in going to the expense of an experiment — affording as it would to others an opportunity to share in its advantage if any.'[74] Four years earlier, when Mushet's process was still in the experimental stage, E. Nevill of the Llanelly Copper Co. refused to take a licence unless it was on 'more advantageous terms than others ... because the first person who uses a new process always works under disadvantage compared with those who follow him'.[75] In practice, the majority of persons taking out a licence usually wanted exclusive use,[76] largely because this gave them almost all the rights of the patentee without any of the attendant problems and costs.

Patentees were generally reluctant to grant exclusive licences unless there was no alternative. Most preferred to license a small number of users, because this enabled them to maximise royalties whilst providing reasonable incentives for the users themselves. Lord Dundonald, who consistently granted semi-exclusive licences for all his patents, is perhaps typical. In 1799 he refused to grant a Mr Case sole use of his patent for white lead because he did not think it was possible for one producer 'to carry on the whole lead manufacture, under my patent, to the extent requisite for the supply of these kingdoms'. He proposed offering four licences, one to Case and the others to the leading firms: 'The present White Lead Manufacturers are very opulent, and of extensive connections, and it would be very imprudent in me, were I not to form a connection with some of the first houses in the Trade, and who have it so much in their power to dam, or injure, at Market the Character of a *New* Article.'[77] Although by the end of the year Dundonald had received a number of applications to use his process, the main firms, who were probably using Turner's alternative process, showed no interest.[78] In the machine-making industry it was also normal for a patentee to license two or three established firms. According to William Jenkinson, no machine maker would bother with a patent if it was open to the whole trade. 'It is hardly worthwhile,' he argued, 'for a machine maker to take up the patent, in as much as there is no patent taken out which does not require expense in fairly introducing it to the public notice, and unless a machine maker has an advantage in the shape of receiving a share with one or two in the patent right, or the exclusive right of building the machine he would wait till the patentee or other persons had introduced the machine'.[79] A restrictive licensing policy was also used by James Hartley in the glass industry, although in this case the patentee was put under some pressure by the two leading firms in the industry. Hartley's patent, which was granted in 1847 for a method of making rolled plate glass, came at a propitious time. The repeal of the glass excise duty in 1845 together with the building boom in the mid-1840s increased the number of glass firms between 1844 and 1847. This intensified competition, and prices began to fall until 1852, forcing over half the firms (mostly those who manufactured crown glass) to leave the industry.[80] In the early 1850s there were ten glass firms, and after Hartley had successfully defended his rights against Hadland (1852) the value of the patent became obvious, especially as the demand for window glass expanded.[81] In December 1854 Hartley licensed the two largest firms, Chance and Pilkington, although rather reluctantly:

> I have no desire to license the patent to any one, it is no advantage to me to receive £500 a year from a House [i.e. £500 per furnace]. I am out of pocket by what you and Messrs. Chance pay me, as compared to what I should be if you did not make rolled glass but it is not good policy for a patentee to grasp too much, but to be satisfied with a fair amount of reasonable competition, hence my satisfaction with out present agreement.[82]

Hartley's 'satisficing' has to be seen in context, though. In the late 1830s and early 1840s many patents were taken out for minor improvements in glass-making to avoid paying royalties to other inventors.[83] By setting a low royalty rate Hartley was at least able to discourage Pilkington and Chance — the more inventive firm technically — from patenting around his patent.[84] His successful lawsuit also made infringment rather risky.[85] Hartley, moreover, was quite willing to license other users — despite his apparent reluctance to do so — and was only dissuaded from doing so by Chance and Pilkington. Together they 'agreed that no other firm should be licensed to use the process without their joint consent', and that they would jointly defend the patent if it were invaded.[86] Between them they managed to control this part of the industry and, by 1860, after acquiring the property of enterprises forced by competition to leave the industry, they were producing 75% of British window glass.[87]

In the glass industry the small number of firms made it easier to operate a restrictive licensing policy. In cotton the chance of using the same tactic was rather more difficult. Arkwright almost certainly attempted to restrict the number of producers operating under his two patents, but he was unable to prevent infringements or competitive patenting. Ten licences in all were granted, and the enormous royalty probably put off others who wished to use the patent legitimately.[88] Robert Bowman, on the other hand, was quite happy to license his improvement in the power loom patented in 1820 to as many users as possible, even though by 1824 only four had taken up the offer.[89] In other branches of the textile trade extensive licensing was common. Kay's invention for spinning wet flax, patented in 1825, was licensed to as many users as were willing to pay,[90] and in the lace industry John Heathcoat ultimately licensed 1,008 machines, nearly 300 of them in 1818.[91] Lister practised a similar policy for his wool-combing machine and after buying rights in J. Heilman's 1846 patent he charged users between £900 and £1,000 per machine, allegedly the highest royalty ever paid.[92] In the sugar refining industry, which in terms of the number of firms was nearer to the glass than the cotton industry, few patentees operated an exclusive or semi-exclusive licensing policy. In his evidence before the 1851 Select Committee Robert Macfie claimed that few inventions were offered for sale and that most inventors preferred to license. Roughly a quarter of the London firms worked under Howard's licence, and Macfie and others used patents taken out by Messrs Terry and Parker, Dr Scoffern and Finzel.[93] In the iron industry Neilson pursued an equally extensive licensing policy and by 1841, thirteen years after his patent was registered, he had licensed sixty-four English and sixteen Scottish ironmasters, and allowed three months for free trials and experiments.[94] John Donaldson also agreed to license any tallow chandler 'so that I am no way connected with the sale or profit and loss' of working the patent.[95] John Howard Kyan's cure for dry rot was marketed in the same way.[96]

Licensing had a number of advantages. Firstly, it reduced the chance of infringement and reduced the transaction costs of negotiating and policing quite separate agreements with individual users. Secondly, it provided patentees with a steady income and created a certain amount of goodwill. It also relieved the inventor from setting up manufacturing on his own account. Inventors, as John Farey argued, were generally eager to license anyone who wanted a licence 'at any sort of fair price', since it demonstrated the value of the patent and served to induce others to apply for the right.[97]

Some users, however, were not always eager to take out a licence because it amounted to recognition of a patent's validity. Heathcoat observed, no doubt from bitter experience, that 'parties do not like to go to a patentee and ask for a licence: [they] generally infringe with a view of destroying the patent'. Asking for a licence, he concluded, 'would go towards acknowledging the validity of ... the patent and as they contemplate that in all probability he would refuse it, it would place them in a worse situation in contesting the patent by giving an opportunity to the patentees of proving that they had thought so well of the validity of the patent as to apply to him for a licence'.[98] Marshalls', who were rarely willing to license any flax machine, refused to give the sanction of their name to an invention patented by Solomon Robinson, despite believing 'there is excellence in the principle'.[99] Licensees, moreover, were not able to recover any rents which had been paid if the patent was later found to be void.[100] When the demands of patentees were reasonable, however, most users would take up a licence, especially if competitors were successfully reducing costs and increasing market shares. In most cases a licence soon paid for itself.[101]

The nature of licence agreements varied as much as the type of licence. Some related the royalty to output (a running royalty), some charged a fixed annual rent, whilst others would set the royalty rate on the number of machines or spindles used. In sugar refining, where profit margins were allegedly small, a running royalty was the favoured method. Howard, who originally sought a rent of 4s per cwt, ultimately charged 1s per cwt, as did Finzel, whilst Terry and Parker charged between 1s 6d and 2s.[102] Neilson also charged 1s per ton for his hot blast[103] and John Donaldson received 4s 8d per cwt for his 'luminators'.[104] James Russell agreed to accept 7½% of the selling price for his patent for welding tubes,[105] and J. C. Fischer offered J. Martineau a licence for new porous bricks for a third of the net profit.[106] The running royalty, however, fluctuated with the state of trade and depended on the efficiency of the user. To reduce these variations in income some patentees fixed the rent to the number of machines or spindles, and thus continued to receive payments during a recession when machines were being used less intensively or not at all. James Kay charged 6s per spindle for his wet spinning process, with an initial payment of £300 to cover the cost of supervising and setting up the first 1,000 spindles.[107] Heathcoat seems to have varied his royalty with the kind of lace produced, but the cost of a licence was generally £20 16s per machine per annum for a medium width.[108] If premiums were not paid, as sometimes occurred in recession, he confiscated the machinery, a practice which Henson called 'pure terror'.[109] In the paper industry Fourdrinier's papermaking machine was licensed in the same way. J. Phipps and John Hayes each paid £100 per annum per vat and, if the 1807 application for an extension of the patent was successful (as it was), it was agreed to reduce premiums.[110] By 1822 forty-two users were licensing the machine, and all were permitted to pay the premium in four annual instalments.[111] Boulton and Watt fixed the premium for steam engines on horse power and charged £5 h.p. per annum in northern counties and £6 in London, because of differences in the price of coal. Watt also exchanged a substantial amount of knowledge and supplied drawings as well as directions for both materials and erection.[112]

Differential royalty rates do not appear to have been common, although patentees sometimes offered preferential licences to some users in the hope of encouraging wide use. John Kyan offered lower rents to the Duke of Buccleuch and some other landowners,[113] but effective price discrimination depended upon confidentiality and separation of markets. Arkwright, as usual, seems to have broken all the rules, and managed a discriminatory policy despite the close location of licensees. Gardum, Pare and Co. paid £2,000 rent for the 1769 patent, and £5,000 for the 1775 one, plus an annual royalty of £1,000,[114] whilst T. Walshamn, another licensee, agreed to pay £4,200 and interest on the annual instalments. In 1786, a year after the patent was set aside, he attempted to avoid the final payment.[115]

Arkwright permitted licensees to use only 1,000 spindles, and such a restriction was often written into agreements. Lord Dundonald's 1795 patent for the manufacture of alkali is, perhaps, typical and illustrates a number of problems some inventors faced when granting licences. In the late eighteenth century alkali was an important input for the developing soap and glass industries, and for certain branches of the textile trade. It was produced mainly from domestic kelp and barilla imported from northern and eastern Europe, both of which were subject to interruptions in supply and violent fluctuations in price. This encouraged the search for a cheaper substitute, and several inventors and chemists sought to produce a synthetic alkali from common salt.[116] Between 1779 and 1789 six patents were taken out for the manufacture of alkali[117] and in the mid-1790s, when Spanish barilla was selling at £60–£70 per ton on the London market,[118] Lord Dundonald took a patent out for England, Ireland and Scotland. In May 1796 he assigned the Scottish patent for an undisclosed but 'valuable consideration', and licensed the English patent separately to Lord Dundas and Aubone Surtees and partners. Dundas was to work the process at Dalmuir (in an old candle works) and Surtees at Bell Close, Newcastle. As centres for the production of glass and soap both districts offered a ready market for soda.[119] The agreement listed the following conditions: Dundonald was to receive 20% of the output each produced, which was reckoned to be equivalent to 33⅓% of the net profit; both licensees were limited to an annual output of 3,000 tons of crystals of soda or alkaline salts for the first three years, and obliged to invest not less than £15,000 in 'appropriate buildings and fit utensils'. Dundonald reserved the right to grant a licence to any glass company to 'make for their own use a certain quantity of crystals' and to award one further licence for his eldest son, Lord

Cochrane, and his three other sons. To prevent over-supplying the market no other licensees were permitted to produce in excess of 3,000 tons per annum, except where a majority vote of all licensees agreed otherwise. In 1798 Dundonald licensed, for £2,000, the British Plate Glass Manufactory, which was then reviving under the management of Robert Sherbourne,[120] 'but restricted them under heavy penalties from making ... alkaline salts for sale or for the purpose of manufacturing any other sort of glass other than that of Plate Glass'.[121]

Dundonald's plan for an 'Alkaline Republic, one and indivisible' hardly got off the ground. Lord Dundas had not produced one 'ounce of goods'[122] by mid-1799, in spite of protestations from Dundonald, who soon wanted to recover the licence to ensure the 'future *character, Credit* and stability' of any undertaking associated with his patent.[123] In one of many letters on the same theme he warned Dundas that it was of 'the *utmost importance* to him and *himself* that projects, as the World calls them, should be *viewed in a favourable point of light*', and reminded him that the 'object of all parties [was] to render the Patents production, as profitable as possible, and as soon as possible'.[124] At Bell Close things were only slightly better, and after two years a mere £4,000 had been invested in the process, and then on the 'erection of works which appertained to the finishing and not to the primary process in making alkaline'.[125] Surtees's failure to meet the output conditions of the licence, despite the valuable assistance of William Losh, led to an acrimonious flow of letters between patentee and licensee, and in early 1798 Surtees began to question the validity of the patent. Dundonald, in turn, threatened to bring an action.[126] By October that year Dundonald had agreed to prolong the licence by one year, fearing that revocation would bring him undeserved discredit.[127] In the following year no progress was made. Surtees continued to impugn the validity of the patent and Dundonald, who had been warned that the 'Newcastle Junta' would infringe his patent, was more and more convinced that legal action was necessary.[128] In the event nothing was done, and by 1801 Surtees withdrew from the industry.[129]

During these years Dundonald, for reasons not associated with his inventions, was deeply embarrassed by financial difficulties which seriously threatened the loss of his ancestral home. The failure of Dundas and Surtees to work the licence also made it more difficult for him to convince others of the value of his invention. Some time in 1798, though, he began negotiating a licence with Benjamin Wood, William Wood and Evan Pugh, London soap makers. Conditions were drawn up and these again limited annual output: 250 tons of crystals of soda, 125 tons of dry carbonate of soda and fifty tons of vegetable alkali or pearl ashes. The production for sale to other users was also prohibited.[130] Dundonald, who had a small alkali works at Poplar, owed William Wood an undisclosed but small sum of money and, as other creditors began calling in Dundonald's debts, Wood began to prevaricate, presumably to reduce the £5,000 which he seems to have agreed to pay for the patent right.[131] Dundonald quickly sensed what was happening and complained against being forced into accepting reduced terms 'with persons who, knowing my situation, might be induced (as is too frequently the case) to take an undue advantage of a Man's necessity'.[132] Dundonald now offered several alternative conditions which relaxed the limits on output and allowed Wood to produce alkali for sale. But as Wood — who was pleased with the experiments he had made — continued to bide his time, Dundonald threatened to sell his patents to the government, and from mid-1799 wrote several letters to the Lords of the Treasury and Lord Liverpool to that effect.[133] In the end Wood refused to take a licence.

In the meantime, Dundas had managed to start manufacturing the alkali in December 1799. William Hamilton, Dundonald's legal adviser, was almost jubilant:

> I found the works in a state of forwardness which I had no conception could have taken place in so short a time ... [and] nothing but want of salts can prevent money coming in before long ... I therefore entreat you to keep up your spirit ... and you must do me at least the justice to say that I never totally dispaired of the Republic ... When the fame of Dalmuir works gets to England the English Patent will, to a certainty, revive and become a Mine of wealth to you and your family ... All I have to recommend to you in the meantime is to keep a sharp look out after an invasion of your Patents.[134]

What happened thereafter is unknown. Clow suggests that Dundas achieved success, but only after abandoning Dundonald's patent for the process of an unnamed inventor who was under a forfeit of £5,000 not to reveal the secret.[135] In all probability a vegetable alkali was being manufactured, as the use of salt did not flourish until the 1820s. Natural soda in the form of kelp and barilla still managed to satisfy demand, despite fluctuations in price, and until the development of the Leblanc process no good artificial soda was available.[136]

Apart from restricting the level of output that licensees could produce, some agreements included a spatial dimension and limited users to producing and selling in certain areas. James Morland, a

Burnley gas tube manufacturer, who agreed to take out a licence for James Whitehouse's patent after having been served an injunction in 1831, could sell his tubes only in Yorkshire, Lancashire, Cheshire, Derbyshire, Lincolnshire, Durham, Northumberland, Cumberland and Westmorland. In these areas he was also prohibited from selling to the Liverpool Coal and Gas Co., Ralph and Spooner, engineers, of Bolton, or to James Bradford, a Manchester manufacturer.[137] By 1838 there were seven other licencees, and these too were probably only permitted to sell their output in other regions.[138] The Marquess of Tweeddale's patent for a machine to make tiles for the drainage of land seems also to have limited the area of sale, as Frederick Tayman was granted a licence to produce tiles for the London district.[139] Beard's licence for the use of his Daguerreotype was restricted by town. Alfred Barber, who took a licence out in September 1841, was permitted to use the invention only in Nottingham and was not allowed to assign the licence without the patentee's consent.[140] How many licence agreements had such restrictions is again unknown, but it seems unlikely that there were many. For most industries, especially those which relied upon the export trade and those with a large number of producers widely dispersed geographically, such restrictions were simply inappropriate. The demand for something like gas tubes was, in contrast, limited to urban areas and could therefore be easily maintained. The policing costs for goods which had a far wider market would have made zoning prohibitively expensive to enforce.

The absence of evidence of cross-licensing on a bilateral or multilateral basis is one rather puzzling feature of the trade in invention, especially for the 1830s and 1840s, when technology was increasingly interrelated. In part this can be explained by the fact that inventors and users were frequently different people, one a seller, the other a buyer of inventive output. It can also be explained, in part, by the number of users who simply infringed to avoid paying any sort of premium. It is also possible that inventors swapped on a patent-for-patent basis without formalising the exchange by contract. There are doubtless other reasons, but until further research is done on this subject the puzzle remains. Licensing, however, was an important method of trading inventions, and the many different types of agreement suggest that inventors thought seriously — at least as seriously as users — how to achieve the best possible return for their output.

Partnerships between inventors and users represented another popular method of disposing of inventions.[141] The economic marriages between Boulton and Watt, Strutt and Arkwright, Marshall,

Benyon and Murray, Macintosh and Hancock, Roberts and Sharpe, Dundonald and Losh are, of course, well known, and indicate that business success during the industrial revolution depended on technical skill as well as entrepreneurial talent.[142] As Matthew Boulton noted in 1788, 'partnership ought to be founded on equitable principles and like a pair of Scales be Balanced either with money, Time, Knowledge, abilities, or possession of a market'.[143] For inventors, partnerships held out a number of advantages. They provided access to greater amounts of capital in order to perfect one or more inventions, and an assured outlet for the product of the invention.[144] Risk was also spread, although capitalists generally carried the greater share and profits were usually divided on that basis. Significantly, partnerships were nearly always formed on the security of patent protection, with the capitalist covering all the expenses of the patent itself.[145] In 1817, for example, John Counter formed a partnership with William Henry Simpson, a mechanic who had invented a new method of spinning wool and cotton with spring levers. Simpson took the patent out in his name but Counter paid all the costs in exchange for a half share in the profits. All other development costs were to be shared equally, and licences could be granted only if both agreed or by two parties chosen by them if they disagreed.[146] Although, like licensing, the nature of partnerships varied, many believed this was the safest way for inventors to profit from their inventions.[147]

Apart from forming partnerships (which will be discussed in more detail in the following chapter), inventors could arrange for others to undertake the sale and licensing of their patents. This does not appear to have been common, and the evidence seems to suggest that it was usually associated with inventors who cared little for manufacturing or partnerships. Stephen Langton, who invented a process for seasoning wood, expected his patent to be very lucrative but would not form a partnership, preferring instead to use an agent to sell the invention. 'This arrangement,' he claimed, 'is intended to obviate the necessity for a partnership or engagement in trade, which, from being a person of landed property, I should have an insuperable objection to forming with any parties however respectable.'[148] The Marquess of Tweeddale arranged the formation of a company to manage the patent for his tile-making machine, but, like Langton, he was not above taking his share of the profits.

During the industrial revolution, then, there was clearly an extensive trade in invention where patentees, through a variety of methods, would attempt to sell their output in the market place. Moreover, the market for invention was quite complex, and although it had no Exchange that could compare with those of the textile trades (although some did suggest establishing something similar) it was fairly well developed. This trade in invention suggests two things. Firstly, whether inventors sold direct to users, used licences, formed partnerships or left arrangements to their agents, their primary if not sole purpose was to make as much profit as possible. Secondly, the trade in invention demonstrates that the relationship between buyers and sellers was *behavioural* rather than *deterministic*.[149] Inventors did not simply respond to user demand. Many inventors did not have a guaranteed market and, as manufacturers were frequently slow to alter the routine of producing goods, had to go out and create a demand for their output.[150]

The last two chapters have attempted to show that the main stimulus to invention and inventive activity during the industrial revolution was profit. That many failed to make any should not disguise the fact that this is what they tried to do. The desire to protect inventive output by patents, the number of multiple patentees, the diversification of invention portfolios and the trade in invention all indicate that the typical inventor was not a harmless bungling genius who occasionally and inadvertently made money from his ideas. It would be absurd to deny that some inventors were cranks and that some were moved by altruism or fame, but few were eccentric enough to be indifferent to the market value of their inventions. If any sense is to be made of invention as a whole, then it should be seen and analysed as an economic activity. In the nineteenth century, at any rate, there was very little disagreement about this issue.

> All who are intimate with the real history of Arts and Sciences and with the views and feelings of those who cultivate them, know well that the hope of individual gain continues as much as ever to be the prime mover in the march of improvement. Of all the more remarkable inventions of modern times there is scarce one which may not be traced to this source, and to the protection to individual enterprise which patents afford.[151]

**Notes**
1. R. Koebner, 'Adam Smith and the industrial revolution', *Economic History Review*, 1959, pp. 381–91.
2. Smith, *op. cit.*, p. 21; see also N. Rosenberg, 'Adam Smith on the division of labour: two views or one?', *Economica*, 1965, pp. 127–39.
3. F. J. Bramwell, *The Expediency of Protection for Invention*, 1875, p. 8.
4. Dircks, *op. cit.*, 1867, p. 77.

Trade in invention    143

5  Carpmael, *op. cit.*, 1832, p. 94; *Select Committee on Patents*, Parl. Papers, XVIII, 1851, pp. 49, 285.
6  *Autobiography*, p. 33.
7  Cf. Jewkes *et al.*, *op. cit.*, pp. 96–7.
8  D. J. Jeremy, *Transatlantic Industrial Revolution: the diffusion of Textile Technologies between Britain and America, 1790–1830s*, Cambridge, Mass., 1981, pp. 54–6. In the seventeenth and eighteenth centuries most patentees were 'outsiders'. Macleod, *op. cit.*, p. 212.
9  J. Coryton, *A Treatise on the Law of Letters Patent*, 1855, p. 22.
10  E. Gauldie, *The Dundee Texile Industry, 1790–1885*, Edinburgh, 1969, p. 139.
11  Langton MSS, Albert to Langton, 'It would be very difficult to sell the process [for seasoning wood] without it being patented and proved by experimenting', 6.12.1825, L.R.O.
12  *Select Committee on Patents*, Parl. Papers, XVIII, 1851, p. 484.
13  *Select Committee on Patents*, Parl. Papers, III, 1829, p. 96; *Select Committee on Patents*, Parl. Papers, XVIII, 1851, pp. 180, 347.
14  *Ibid.*, pp. 253–4.
15  Dircks, *op. cit.*, 1867, pp. 75–6.
16  Macfie, *op. cit.*, 1863, p. 20. Macfie suggests that selling inventions was one of the commonest methods used by inventors.
17  These figures have been weighted downwards slightly because where there is no information to indicate whether a patent was sold, it has been treated as if it was *not* sold. If unknown cases are excluded, then the proportion of patents sold increases as follows: 1770–99, 8%; 1800–29, 21%; 1830–39, 25%; 1840–49, 25%.
18  *The Commissioners of Patents' Journal*, 1859, pp. 446–63.
19  E. Holroyd, *A Practical Treatise on the Law of Patents for Invention*, 1830, p. 156, n. L.
20  Ladd. 701, W. Collins to Messrs Cox and Son, 12.3.1808, Birmingham University Library.
21  Bessemer, *op. cit.*, p. 37.
22  *Ibid.*, pp. 59, 81.
23  T. C. Barker, *The Glassmakers: Pilkington, 1826–1976*, 1977, pp. 440–1.
24  W. G. Rimmer, *Marshall's of Leeds: Flax Spinners 1788–1886*, Cambridge, 1960, pp. 265–6. The rights were ultimately bought by Lister, who then appears to have given Marshall an exclusive licence for combing flax.
25  Mushet MSS, 17.11.1838, D646/142/1–5, Gloucestershire County Record Office (hereafter G.C.R.O.).
26  Coryton, *op. cit.*, p. 143.
27  *Mortimer's London Directory*, 1763, 'Though pins are so necessary ... there are but few makers, the cheapness of the commodity from which only a small profit arises, affording very little encouragement for new beginners.' Seven tradesmen were listed as pin manufacturers, but there certainly include merchants as well as manufacturers. See also S. R. H. Jones, 'Price associations and competition in the British pin industry, 1814–1840', *Economic History Review*, 1973, pp. 237–53.
28  T. S. Ashton, 'The records of a pin manufactory', *Economica*, 1925, pp.

281–92; see also C. Babbage, *On the Economy of Machinery and Manufactures*, 1832; C. F. Pratten, 'The manufacture of pins', *Journal of Economic Literature*, 1980, pp. 93–6.
29 *Select Committee on Employment of Children*, Parl. Papers, XV, 1843, p. 26.
30 Pointing had experienced little technological change over two centuries. The replacement of the file by the steel mill was the only advance in this area.
31 Patent No. 2182.
32 Patent No. 3268. It seems that Harris was a City hosier in the early 1790s, and then a cotton spinner in Nottingham some three or four years later. When he went bankrupt in 1797 he still owned a City hosiery warehouse and a wire mill at Waltham Abbey. He therefore had contact with pin makers as a supplier of their major input. S. D. Chapman, *The Early Factory Masters*, Newton Abbot, 1967, pp. 22, 81, 141–2, 231.
33 Woodcroft, *Index ..., op. cit.*, p. 81.
34 *Rees Encyclopaedia*, XXVI, 1819, see article on pins; Gutch MSS, Hall to English, 12.10.1824, Redditch Public Library (hereafter R.P.L.).
35 *Patent Abridgement Specifications, Pins and Needles*, p. 3.
36 Woodcroft, *Index ..., op. cit.*, p. 62.
37 *Select Committee on Mr. Bradbury's Petition relative to Machinery for Engraving and Etching*, Parl. Papers, III, 1818, pp. 361–2.
38 Patent No. 4129. The 1817 patent was a communication to Hunt, which suggests he may not have been the inventor. In 1814 an American, Moses Morse, had patented a completely automatic machine in the United States to which Hunt's machine seems to bear a resemblance.
39 Stubs Wood MSS, Hunt to Stubs Wood, 3.2.1817; Stubs Wood to Hunt 23.2.1817, Letter Book, M.R.L.
40 It was subsequently alleged that Wright pirated a number of Hunt's ideas, an argument advanced by Cowcher's when brought to court on a charge of infringement by Marling, an assignee of Wright's 1824 patent. *Marling v. Kirby*, 1841.
41 Woodcroft, *Index ..., op. cit.*, p. 640.
42 Gutch MSS, Gutch to English, 6.5.1826, B.A., 3586/4, Worcester County Record Office (hereafter W.C.R.O.).
43 Gutch MSS, Hall to English, 11.6.1822, 26.6.1822, 10.7.1822, Box 9, R.P.L.
44 Stubs Wood MSS, J. Wood to Stubs Wood, 25.7.1823, M.R.L.
45 Webster, *Reports*, I, pp. 575–7.
46 Gutch MSS, Gutch to English, 6.5.1826, 26.2.1829; Hall to English, 27.12.1829, B.A. 3586/4, W.C.R.O.
47 Carpmael, *Reports*, pp. 517, 519.
48 Gutch MSS, Hall to English, 16.3.1834, Gutch to English, 13.5.1834, Box 15, R.P.L.
49 Gutch MSS, Hall to English, 16.3.1832, Box 18, 8.10.1834, Box 16; Gutch to English, 29.9.1834, Box 15, R.P.L.
50 Gutch MSS, Hall to English, 16.3.1832, 12.5.1832; Gutch to English, 2.8.1833, 9.6.1834, Box 15, R.P.L.
51 Prosser, *op. cit.*, p. 79.

52 R. C. O. Matthews, *A Study in Trade Cycle History: Economic Fluctuations in Great Britain, 1833–1842*, Cambridge, 1954, pp. 43–68.
53 Patent No. 6578, 1834. He also took out a similar patent in 1835, Patent No. 6911.
54 Patent No. 6577 and 6768.
55 Gutch MSS, Barnet to English, 29.9.1842, Letter Book R.P.L. After 1842 the American pin industry operated behind substantial tariff protection. T. Bate to English, 14.9.1842, B.A. 3586/9, W.C.R.O.
56 This section is based on J. L. Bishop, *A History of American Manufactures, 1608–1860*, Philadelphia, 1868, 3rd ed.
57 For a more general discussion of this problem see R. R. Nelson and S. G. Winter, 'Neoclassical vs. evolutionary theories of economic growth: critique and prospectus', *Economic Journal*, 1974, pp. 886–905.
58 Barlow, *op. cit.*, pp. 17–18; Dircks, *op. cit.*, 1867, p. 164.
59 *Select Committee on Patents*, Parl. Papers, XVIII, 1851, p. 199.
60 *Loc. cit.* There is no evidence to indicate whether Price and Co. brought an action.
61 *Loc. cit.*
62 *Ibid.*, p. 249; *Select Committee on Patents*, Parl. Papers, XXIX, 1864, p. 107. See also J. Kieve, *The Electric Telegraph: a Social and Economic History*, Newton Abbot, 1973.
63 *Select Committee on Patents*, Parl. Papers, XVIII, 1851, p. 393; Webster, *op. cit.*, 1853, pp. 23–4.
64 Spencer Moulton MSS, J. Clark to S. Moulton, 21.7.1849; J. A. Jacque to S. Moulton, 21.12.1850; W. Woodruff, *The Rise of the British Rubber Industry in the Nineteenth Century*, Liverpool, 1958, pp. 133–44; Payne, *op. cit.*, p. 14. It is important to note that both firms were also approached by inventors.
65 Payne, *op. cit.*, p. 23.
66 Bessemer, *op. cit.*, p. 115.
67 *Select Committee on Patents*, Parl. Papers, XXIX, 1864, p. 429.
68 *Select Committee on Patents*, Parl. Papers, XVIII, 1851, p. 100.
69 For a general and recent discussion see J. A. D. Cropp, D. S. Harris and E. S. Stern, *Trade in Innovation*, 1970, especially ch. 4. See also Taylor and Silberston, *op. cit.*, pp. 113–24, 162–6, 181–6.
70 *Select Committee on Patents*, Parl. Papers, XVIII, 1851, p. 431.
71 PRO/LO/1/15. The patent was taken out in 1840 and an improvement was patented in 1841. In 1850 the firm bought the patent for £855.
72 MSS. 12.5.1830, H.L.R.O.
73 P.R.O. BT/1/478, 29.6.1850.
74 Mushet MSS, R. Hopkins to D. Mushet, 7.3.1839, D. 2646/14J. G.C.R.O. In Wales bar iron was made by putting the pigs in the puddling furnace and mixing them with cinders. Mushet's process used ore rather than cinders.
75 Mushet MSS, E. Nevill to D. Mushet, 4.11.1835, D 2646/135, G.C.R.O.
76 Prosser, *op. cit.*, p. 552.
77 Dundonald MSS, Dundonald to Case, 22.6.1799, GD233/109/7/F.12, S.R.O.
78 Dundonald MSS, Dundonald to Mr Wilkinson, 17.11.1799, GD/233/109/

7/17. Dundonald was at this time having difficulties with licensees of his alkaline patent and began to question 'the propriety or Prudence of granting licences to any persons' for his white lead patent. See GD233/109/7/13, 24.7.1799, unaddressed letter, S.R.O.
79 *Select Committee on the Export of Machinery*, Parl. Papers, VII, 1841, p. 120.
80 Barker, *op. cit.*, 1977, p. 99.
81 *Hartley v. Hadland*, 1852.
82 Cf. Barker, *op. cit.*, 1977, p. 103.
83 In the 1830s Pilkingtons' infringed Chance's patent for the process of polishing and smoothing thin sheets of glass known as patent plate glass. Although this finishing machinery could be used only under licence, Chances' never brought an action. T. C. Barker and J. R. Harris, *A Merseyside Town in the Industrial Revolution: St. Helens, 1750–1900*, London, 1959, pp. 213–15.
84 In the 1852 court case it was alleged that Chances' had invested £25,000 in perfecting the process, without success.
85 Chances' did actually infringe the patent, but probably before the Hadland case. Barker, *op. cit.*, 1977, p. 103.
86 *Ibid.*, p. 104.
87 *Ibid.*, p. 107.
88 Chapman, *op. cit.*, 1967, pp. 73, 75, 128, and Chapman, *op. cit.*, 1965, p. 533.
89 *Select Committee on Arts and Manufactures*, Parl. Papers, I, 1824, p. 309.
90 Marshall MSS, W. Renshaw to Marshall, 29.7.1831, E/16/17, B.L.
91 Felkin, *op. cit.*, p. 250.
92 J. Burnley, *The History of Wool and Woolcombing*, 1889, p. 240. *Select Committee on Patents*, Parl. Papers, XXIX, 1864, p. 532.
93 *Select Committee on Patents*, Parl. Papers, XVIII, 1851, p. 150; *Select Committee on Patents*, Parl. Papers XXIX, 1864, pp. 116, 465–7, 146–7; *Hansard*, CXVIII, 1851, p. 1851. It is interesting to note that Macfie managed to negotiate a lower royalty for the Howard's patent because he agreed terms at an early stage.
94 *Neilson v. Harford*, 1841, p. 109, B.M.
95 Boulton and Watt MSS, J. Donaldson to Boulton 28.11.1787, letter no. 140, Birmingham Assay Office.
96 Dartmouth MSS, D 564/11. Staffordshire County Record Office (hereafter S.C.R.O.).
97 *Select Committee on Patents*, Parl. Papers, III, 1829, p. 143. See also the evidence of J. Platt, *Select Committee on Patents*, Parl. Papers XXIX, 1864, p. 400.
98 *Lords Select Committee on Patents Amendment Bill*, 1835, H.L.R.O.
99 Marshall MSS, J. Marshall Jnr. to J. Marshall, 11.11.1827, E/17/17 B.L.
100 *Hayne v. Maltby*, 1789; *Chante v. Leese*, 1838, cf. Hindmarch, *op. cit.*, p. 69.
101 The relationship between inventor and licensee deserves further study. The purpose here, though, is simply to show that inventors were concerned primarily with attempting to maximise the return to their inventive effort.

## Trade in invention    147

102 *Select Committee on Patents*, Parl. Papers, XVIII, 1851, p. 116; *Select Committee on Patents*, Parl. Papers, XXIX, 1864, pp. 119, 146, 150.
103 *Neilson* v. *Harford, op. cit.*
104 Boulton and Watt MSS. Donaldson to Boulton, 28.11.1787, letter No. 140, Birmingham Assay Office.
105 Russell MSS, 67/14/22, Bundle 16, S.C.R.O.
106 Henderson, *op. cit.*, p. 28.
107 Marshall MSS, W. Renshaw to J. Marshall, 29.7.1831, E/16/17 B.L.
108 *Select Committee on Arts and Manufactures*, Parl. Papers, I, 1824, p. 373.
109 Add. MSS, 27,807, Henson to Place, 31.5.1825.
110 *Minutes of the Proceedings of the Committee of the House of Lords*, Parl. Papers, XIV, 1807, pp. 16, 17, 27. It appears that Hayes had made an agreement which allowed him free use if the patent was extended, as it was by fifteen years. 47 George III, cap. 131.
111 D. C. Coleman, *The British Paper Industry, 1495–1860*, Oxford, 1958, p. 232.
112 Hartree MSS, J. Watt to McGrigor, 30.10.1784, Devon County Record Office; Boulton and Watt MSS, Watt to C. J. Eyre, March 1795; see also a rough draft, 27.11.1795, Parcel E, B.R.L.
113 Tweeddale MSS, R. Brown to Marquess of Tweeddale, 27.11.1835, Scottish National Library (hereafter S.N.L.).
114 Chapman, *op. cit.*, 1967, p. 73.
115 Strutt MSS, T. Walshman to Strutt senjor, 4.7.1786, Derby Public Library (hereafter D.P.L.).
116 For a discussion of these early experiments see A. and N. L. Clow 'Vitriol in the industrial revolution', *Economic History Review*, 1945, pp. 45–55; Musson and Robinson, *op. cit.*, 1969, pp. 352–71; R. E. Schofield, *The Lunar Society of Birmingham*, Oxford, 1964, pp. 66–7, 76–9.
117 Musson and Robinson, *op. cit.*, 1969, p. 364. It is interesting to note that James Watt did not take out a patent for the manufacture of alkali because of the duties on salt, 'and the disagreeable circumstances of being attended by excise Officers together with the moderate price of alkaline salts arising from the plentiful [*sic*] importation of American Potashes'. Watt to Keir, 3.5.1780, cf. Musson and Robinson, *ibid.*, p. 361.
118 Dundonald MSS, GD233/107/e/10, Nov. 1797, S.R.O.
119 Clow, *op. cit.*, 1945, p. 159; Barker, *op. cit.*, 1977, p. 17.
120 Barker, *ibid.*, 1977, pp. 20–2.
121 Dundonald MSS, GD233/109/7/2 undated; see also Dundonald to W. Wood, 19.5.1799, GD233/110/J/19; and GD233/109/7/F.3, undated, GD233/110/K/24, S.R.O.
122 Dundonald MSS, W. Hamilton to Dundonald, 7.5.1799, GD233/110/K/78, S.R.O.
123 Dundonald MSS, GD233/110/K/24, undated, S.R.O.
124 Dundonald MSS, GD233/110/K/25, undated, S.R.O.
125 Dundonald MSS, Dundonald to J. Surtees, undated, GD233/110/I/2/1 and I/2/3. See also GD233/109/7/F.3, S.R.O.

126 Dundonald MSS, Dundonald to A. Surtees, Jan. 1798, GD233/110/I/18, S.R.O.
127 Dundonald MSS, Dundonald to Surtees, 16.10.1798, GD233/110/I/18; Dundonald to Surtees, 4.11.1798, GD233/110/I/21 and I/23, undated, S.R.O.
128 Dundonald MSS, Dundonald to Surtees, 5.1.1799, GD233/110/I/25; Dundonald to Surtees, 21.5.1799, I/28; and I/1.31(2), undated; Dundonald to Dundas, 11.5.1799, GD233/110/K/88, S.R.O.
129 Dundonald MSS, 4.5.1801, unaddressed, GD233/110/I/34, S.R.O. For a discussion of Dundonald's activities in Newcastle see A. Clow and N. Clow, *The Chemical Revolution*, 1952, pp. 100–07.
130 Dundonald MSS, GD233/100/J/1, undated, and 100/J/10, also undated, S.R.O.
131 Since nearly all the letters are undated and some missing, it is difficult to know whether both parties actually agreed on this sum of money.
132 Dundonald MSS, Dundonald to W. Wood, 1.5.1799, GD233/100/J/14; see also J/25, undated, S.R.O.
133 Dundonald MSS, Dundonald to Lords of Treasury, 15.7.1799, GD233/109/7/9; Dundonald to Lord Liverpool, undated, 107/E/12 and 107/E/3, undated, S.R.O.
134 Dundonald MSS, W. Hamilton to Dundonald, 2.12.1799, GD233/110/K/98, S.R.O.
135 Clow, *op. cit.*, 1945, p. 160.
136 For a discussion of the delay in the development of manufacturing alkali by common salt see D. W. F. Hardie, 'The Macintoshes and the origins of the chemical industry', in Musson, *op. cit.*, 1972, pp. 168–72; Musson and Robinson, *op. cit.*, 1969, p. 368; T. C. Barker, R. Dickinson and D. W. F. Hardie, 'The origins of the synthetic alkali industry in Britain', *Economica*, 1956, pp. 158–69.
137 Russell MSS, 67/14/22, Bundle No. 16, S.C.R.O.
138 J. F. Ede, *History of Wednesbury*, 1962, p. 237.
139 Tweeddale MSS, S. G. Birnie to Tweeddale, 19.4.1839, S.N.L.
140 *Beard* v. *Barber*, 1843. Beard, who took out two patents in 1842 and 1840, appears to have been in the textile trade.
141 *Select Committee on Patents*, Parl. Papers, XVIII, 1851, p. 272.
142 See N. McKendrick, 'Wedgwood and Thomas Bentley: an inventor-entrepreneur partnership in the industrial revolution', *Transactions of the Royal Historical Society*, 1964, pp. 1–34.
143 Cf. E. Robinson, 'The early diffusion of steam power', *Journal of Economic History*, 1974, p. 105.
144 *Lords Select Committee on Patents Amendment Bill*, 1835, H.L.R.O. Rosser believed that inventors formed partnerships because they were 'likely to continue in connection with the invention [and] like the natural parent of a child, being best to nurture his own offspring — more fitted than a foster parent, to bring it to healthy maturity' (unpaginated).
145 *Select Committee on Patents*, Parl. Papers, XVIII, 1851, pp. 32, 414.

146 Mason Tucker, MSS, Articles of Agreement, 9.3.1817, 924M/B8/4, Devon County Record Office.
147 See, for example, Webster, *op. cit.*, 1853, p. 26.
148 Langton MSS, Langton to John Turnley and Sons, 17.12.1827, L.R.O.
149 Nelson and Winter, *op. cit.*, pp. 886–905.
150 This is what entrepreneurs did in other spheres. See N. McKendrick, 'Josiah Wedgwood: an eighteenth-century entrepreneur in salesmanship and marketing techniques', *Economic History Review*, 1960, pp. 408–33; E. Robinson, 'Eighteenth-century commerce and fashion: Matthew Boulton's marketing techniques', *Economic History Review*, 1963, pp. 39–60; J. Tann, 'The international diffusion of the Watt engine, 1775–1825', *Economic History Review*, 1978, pp. 541–64, and 'Marketing methods in the international steam engine market: the case of Boulton and Watt', *Journal of Economic History*, 1978, pp. 363–91.
151 *Mechanics Magazine*, LV, 1851, p. 34. See also Barlow, *op. cit.*, pp. 66–7 and Dircks, *op. cit.*, 1867, p. 75.

# 8

# Investment in patents

The industrial revolution was an affair of economics as well as of technology. The growth of savings and the willingness to invest in industry was crucially important and 'made it possible for Britain to reap the harvest of her ingenuity'.[1] Technological change would have been seriously limited had not considerable capital resources been invested to develop and exploit inventions.[2] Landes emphasises the relative ease with which inventors were able to secure financial backing for their projects.[3] There has not, however, been any analysis of the cost of producing and developing inventive output, nor has there been any analysis of the role which patents played in protecting and inducing investment in invention.[4] The purpose of this chapter is to come to grips with this problem.

The reason why it has been ignored is simple enough: the evidence is almost non-existent. Inventors, innovators and capitalists rarely kept detailed records of the cost of producing and developing inventive output, largely because there was no urgent reason to do so. Those who produced and used new techniques generally enjoyed a strong market position, and consequently had no need for an elaborate system of accounting to prove that the invention was worth adopting. Intuition and a feel for the market appear to have been enough. Indeed, it was 'precisely at the time when firms departed from the routine and innovated, that accounting was neglected ...'.[5] Few inventors, it seems, ever thought in terms of columns of debits and credits. Even when records exist, the practice of mixing different business accounts is another important factor affecting the quality of the evidence. Henry Reeve, registrar to the Judicial Committee of the Privy Council, which examined applications for extending the life of patents,[6] claimed that inventors 'may be men of business who are engaged in other trades, perhaps as manufacturers, and the accounts which strictly appertain to the patent are mixed up with the accounts of the other business,

## Investment in patents 151

whatever it may be ... [It] is exceedingly difficult ... to discover how much appertains to the invention and how much appertains to their ordinary pursuits in business.'[7] On the rare occasions when separate accounts were kept, it is often impossible to distinguish between the costs of inventing and of innovating, or to assess how typical they are. Inventions, almost by definition, are unique. That some inventions were more costly to produce and develop than others is perhaps the only generalisation which can be made with any certainty.

According to Landes, one significant feature of the industrial revolution was 'the relative ease with which inventors found finance for their projects ... [and] ... the rapidity with which the products of their ingenuity found favour with the manufacturing community'.[8] This argument seems to be strongly influenced by the recent literature which stresses the relative abundance of capital during the industrial revolution, and needs to be qualified in a number of ways. Firstly, investment in inventive output depended upon the perceived value of the invention itself (which means mistakes could be made). Few capitalists would waste their resources on a project which they believed was unlikely to reap an adequate rate of return. When William Carpmael claimed that he 'knew of no invention ... which wants capital to carry it out', he was specifically referring to inventions of value.[9]

Secondly, few capitalists would invest in invention without the protection of a patent. Inventing was a risky activity and this kind of protection was the only realistic way of appropriating a return sufficient to cover the cost of producing and developing inventive output. R. H. Wyatt, editor of the *Repertory of Arts*, believed that many capitalists invested their money 'in *important* patents in consequence of protection',[10] and W. A. Cochrane considered that 'capitalists would not embark on a speculation that was not protected'.[11] Bessemer claimed it would be 'sheer madness'[12] for men to waste their energy and money without the protection of a patent.[13] And there is very little doubt that Boulton would not have advanced the capital to finance Watt's steam engine without the security of the 1775 Act, which extended the patent for an unprecedented period of twenty-five years.[14] Boulton needed to be reassured that his investment would yield an adequate profit:

> But as a great part of the time of his patent is elapsed and his [Watt's] own life very precarious, and as a large sum of money must yet be expended before any advantage can be gained from it, I think that his abilities and

my money may be otherwise better employed, unless Parliament be pleased to grant a prolongation of the term of the exclusive privilege.[15]

Investment in inventive output was also limited, although less seriously, by a third factor. In each patent there was a proviso which limited to five the number of persons who could have a share in the patent. If the number exceeded the limit the patent was automatically void. This clause, which was introduced to conform with the Bubble Act of 1720, was designed to prevent large legal monopolies acting against the public interest, and to eliminate any possibility of a 'jobbing' market developing in patent shares.[16] A few inventors mistakenly thought that the proviso restricted the number of licences which could be granted,[17] but a larger number complained that it discouraged capitalists from investing in inventions. Several attempts were made to alter the restriction, and nearly every Bill for reform sought to extend the number who could be actively involved in the promotion of a patent.[18] Even though the Bubble Act was repealed in 1825 the restriction continued in force until 1832, when the Attorney General, Sir T. Denman, after having been approached by William Carpmael, agreed with Lord Auckland, President of the Board of Trade, to extent the number of twelve.[19] Inventors could, of course, form unincorporated companies by private Act of Parliament, to raise capital in larger amounts than could be obtained through partnerships, but hardly any took up the option until after 1837, when the Board of Trade could, at its discretion, confer by charter or letter patent full corporate status, including the privilege of limited liability.[20]

In the early 1850s, when the demand for limited liability was suddenly stepped up, a number of people began to argue that the risk of investing in invention was too high without some additional form of protection. In the 1851 Select Committee on the law of partnership this view was put forward by several witnesses, some of whom also gave evidence to the committee on patents. The solicitor, John Duncan, said that the 'chief oppression that a patentee is exposed to, is not the high cost of patents but the operation of the partnership laws ... which can scarcely offer security to a creditor'.[21] This theme was repeated by a number of others, and Babbage even claimed 'that many an excellent invention is lost for want of capital to carry it out'.[22] There is, though, no evidence to support the contention that limited liability would have helped inventors in any real way. In fact, during the industrial revolution the issue of limited liability rarely emerged as an issue, and it probably came out into the open in 1851 only because patent law was being considered at the same time as the law of partnership. The limit on the number of parties who could invest in inventions was a far more critical point, although this, too, does not seem to have deterred inventors to anything like the extent that was claimed.[23]

Two other rather more general constraints on investment need to be considered: firstly, the difficulties of raising capital at different periods over the trade cycle, and secondly, the effects of the changing rate of profits. Investment is a discontinuous process and fluctuates more violently than current production. Large changes in investment usually occurred, especially in the first half of the nineteenth century, in the latter part of the expansion phase after excess capacity had been eliminated. Since patent activity fluctuated with the trade cycle — as will be shown in the next chapter — it is quite possible that inventors experienced rather more difficulty finding investment funds in the downswing than they did in periods of boom. But whether investment in invention took place during recessions, when the possibility of cost reductions increased the chance of improving the position of user firms, or in booms, when profits were available, is unclear. Crouzet suggests that the major cycles in investment, in 1783–92 and 1825–36, were largely connected in the first case with the adoption of Arkwright's, Crompton's, Cort's and Watt's inventions, and in the second with the spread of the power loom, the self-acting mule, the hot-blast process and railways, but here the investment is in the diffusion of a piece of technology, not in invention or innovation.[24] At present there is no evidence to indicate how investment in inventive output varied over the trade cycle, but it is probably safe to conclude that the volume of this kind of investment increased in booms. Expectations were optimistic, and funds (irrespective of marginal changes in the rate of interest) were generally available.[25] Whether this made it easier for individual inventors, who relied on outside sources of investment, is quite another matter, especially if patent activity outstripped investible funds. Competition between inventors for financial backing may well have been more intense at the peak of the trade cycle than during periods of recession.

The changing rate of profit is the final constraint on investment in invention, and this depends upon the relative supplies (and elasticities) of co-operative factors of production. According to Habakkuk, if the rate of capital accumulation is faster than the growth of the labour force, this leads to increases in real wages and to bottlenecks in production which, in turn, provide incentives for the adoption of labour-saving and capital-using techniques. Moreover, the need to avoid a 'falling rate of profit (caused by the relative scarcity of labour) is a more powerful incentive to devise new methods than the possibility of increasing the rate of profit'.[26] Using this analysis, Habakkuk divides the industrial revolution into two broad periods. Prior to 1800, when labour was relatively scarce and the rate of profit falling, users had a strong incentive to innovate and invest in inventive output. After 1815 the supply of labour was more elastic and there were, consequently, fewer constraints on the rate of profit and hence capital accumulation. This did not slow down technical progress, but users were largely concerned with the wider diffusion of *existing* innovations rather than with the adoption of inventions, because a relative 'capital shortage is less likely to be a continuous stimulus to innovation than a labour shortage'.[27] Since patent activity increased (absolutely) after 1815, the implication is that proportionately less funds were available for financing the production and development of inventions or, at best, were directed to those with a specific and obvious labour-saving content, when short-run changes (e.g. strikes) dictated. This aspect of Habakkuk's analysis does at least lend some support to the argument that investment in invention was not so easy as Landes alleged.

Historical sources are usually better at recording success than failure, but there is some evidence to show that inventors had problems in finding financial support. Lord Dundonald's letter to John Retrie perhaps typifies the feelings of inventors:

> I lately applied to a person acquainted with monied men to know if he would recommend me a person (not terrified at the word projectors or project) who would support in the monied way certain discoveries or undertakings *provided* they should appear to him on a *thorough enquiry* and *investigation* deserving of it ... the patent and the manufacture, if required be vested in the person who advanced the money.[28]

Despite Dundonald's endearing honesty, Retrie, who was mentioned as a likely source of funds, showed no interest. Samuel Clegg's gas meter patent met with similar opposition from most gas companies in the country,[29] and the lack of support for Perkin's patent for central heating compelled him to 'devote himself to the superintendence and introduction of the invention'.[30] George Fredrick Muntz, Daniel Stafford, Parkes and others related similar stories to the Judicial Committee of the Privy Council when applying for patent extensions.[31] This, of course, was the kind of evidence likely to influence the council, but convincing investors of the value of an invention was rather more difficult than has been suggested. As noted earlier, the mass of users were often reluctant to break with routine, unless it could be demonstrated in a fairly obvious way what the benefits were. Investors, knowing that most inventions were unprofitable, needed to be assured of the likely flow of returns, and the likely costs of producing and developing the invention. Inventors, on the other hand, were generally unable to provide this kind of assurance, simply because the outcomes of inventing are necessarily uncertain. Some investors were as sanguine as inventors but many were not, and this explains the paradox noted by a number of informed observers, 'that inventors' patrons are among their most inveterate opponents'.[32] To overcome this problem inventors had to market their inventions with care and do everything within their power to avoid being branded as mere projectors.[33] Some were successful, others failed,[34] but in the end investment in invention depended on the expectation of its value and on patent protection, and in that order.

The most important sources for investment in the production and development of patented inventions are the applications made by patentees for prolongation of their patents. Prior to 1835 extensions were obtained by a private Act of Parliament, and in this period approximately twenty-five applications were considered by various Select Committees. After Lord Brougham's 1835 Act renewal applications were examined by the Judicial Committee of the Privy Council and, down to 1852, seventy-seven cases were considered. The cost of patent extensions, which varied between £350 and £700, together with the reluctance of Parliament and the Judicial Committee of the Privy Council to extend patents unless a very clear case was made, explains why only 100 or so inventors bothered to apply.[35] But, apart from the problem of small samples, the evidence is deficient in other respects. Firstly, applications prior to 1835 are generally recorded in the *Journal of the House of Commons*, and only brief reports are printed, many of which do not mention the costs of developing inventions. The reports of the Judicial Committee of the Privy Council are rather better, but are based on notes published by William Carpmael and Thomas Webster, since the originals were destroyed in 1940.[36] Some of their reports are very useful but many, again, do not reveal the amount of investment in invention. Secondly, and more serious, the records frequently fail to distinguish between the costs of invention, innovation and other aspects of manufacturing.[37] Thirdly, since patents were extended only if they were beneficial to the public and had demonstrably failed to reward the patentee,[38] the sample is probably biased in favour of inventions which were relatively expensive to produce and develop. Finally, there is very little doubt that inventors always made it appear that returns were low; as the Lord Advocate warned, some 'might keep out of view some profits and exaggerate the expenses'.[39] In all probability this explains why petitions did not always separate patent expenses from other costs.[40]

To put what follows in some perspective, it is perhaps useful to examine briefly how much finance and capital was required by industry generally. In the eighteenth century few industries needed much in the way of fixed capital and most firms relied upon a complex network of credit to finance work in progress. In cotton, during the 1790s, a large mill employing 1,000 spindles cost roughly £3,000, whilst one with twice that capacity was £5,000.[41] In the early nineteenth century capital requirements were not significantly different.[42] With the rise of the integrated firm between 1825 and 1850, the amount of fixed capital increased, and in the 1830s a typical mill employing an average of 400 operatives cost somewhere in the region of £20,000 to £50,000.[43] In iron, which by 1800 was the most capital-intensive industry, investment in fixed capital was typically some £20,000 and working capital requirements were fairly similar.[44] In the 1830s small metal trades such as nailmaking, gunmaking and needle making were less capital-intensive. The average capital for a nailmaker was £6,875, with a turnover of £17,500 per annum, whilst toy makers and needle makers had an average capital of £10,400, with an average turnover of £21,250.[45] All this confirms, however, that investment was not excessively large during the industrial revolution, especially before the 1830s. In volume, working capital was certainly far more important. Chapman estimates that even as late as 1842 Horrockses, Miller and Co., Preston's largest cotton manufacturers and one of the biggest firms in Lancashire, had a ratio of fixed to working capital of only 1:1·5, whilst Cottrell suggests that the ratio for secondary metal trades was somewhere between 1:3 and 1:2.[46]

This distinction is less easy to make for investment in invention, but in some cases substantial sums were involved. The Fourdrinier papermaking machine, which was patented in England by John Gamble in 1801 (an improvement was also patented in 1803), cost £60,228 to develop between 1801 und 1807.[47] The invention itself was bought from Didot for a sum of £14,879 and Bryan Donkin, the engineer employed to improve the machine, used £31,667 provided by Fourdrinier to perfect the method. Between June 1802 and June 1806 upward of £11,900 was spent on experiments at mills which Fourdrinier had opened at St Neots, Huntingdonshire, and at

Frogmore and Two Waters, Hertfordshire. In the early part of the nineteenth century a paper mill usually cost around £7,000 and Fourdrinier's St Neots mill, which cost £11,276, was exceptional.[48] The cost of experiment was, therefore, very high and represents over one third of the total amount invested in machines and plant. In the metal trades, investment in invention could be equally costly. Russell, who assigned Whitehouse's patent for pipe tubes, spent £7,429 by 1831 and by 1834 this had increased to £11,000. Francis Bramah, who inspected the plant, estimated that some £10,000 to £12,000 had been sunk in machinery, buildings and power, and reckoned that if the patent was extended the property would have a market value of £20,000.[49] A. M. Perkin's 1831 patent for a primitive form of central heating, which was an adaptation of Whitehouse's invention, invested £47,282 in materials, labour and tools, and £2,700 in experiments and his patent. When the books were examined 'other matters [were] mixed up with the patent business', but the cost of producing and developing the invention were still considerable and the patent was extended for a further five years.[50] The evidence relating to Porter's 1838 patent for an improved anchor suffers from the same problem. Between 1841 and 1849 the accountant E. M. Smith estimated that £78,257 had been paid out to develop and manufacture the article, which all the most 'respectable houses' had refused to produce.[51] T. Jones's 1826 patent for iron wheels, which was assigned to Riddle and Piper, was less costly, but between 1833 and 1840 some £7,284 was invested in machinery patterns, tools, forges and premises.[52] Joseph Bramah's new lock, patented in 1784, cost £10,000 to develop,[53] whilst Cort was alleged to have spent £20,000 on his inventions.[54]

In textiles similar sums were expended, although less frequently. Richard Dossie, of the Society of Arts, suggested (with some exaggeration) that by 1768 Wyatt and Paul had spent £60,000–£70,000 to establish the use of their spinning machines.[55] Arkwright claimed that he had invested £13,000 by 1774, and further evidence indicates that another £27,000 was spent over the next eight years, although much of this was obviously associated with the building of mills.[56] Edmund Cartwright's clerk, John Pasman, estimated that £9,000 had been spent on the power loom prior to 1790 and upwards of £5,000 between 1790 and 1793.[57] In 1838 Richard Roberts noted that his 1825 patent for the self-acting spinning mule cost £29,944 for wages, materials plus an allowance for interest and his time.[58] Since textile machinery was fairly simply and inexpensive, it seems likely that the bulk of inventions cost less to develop. Robert Bowman's patent for an improved power loom cost him £4,000,[59] and William Southworth's apparatus for facilitating the drying of calicoes, muslins and other fabrics cost between £1,500 and £2,000, whilst Kay spent only approximately £1,000 in perfecting his flax wet spinning process.[60]

The cost of inventions for motive power also varied considerably. Six years after Watt's 1769 patent £10,000 had been invested, in spite of the fact that the capital required in the engine trade was usually modest.[61] A. Rosser alleged that one of his clients spent £30,000 in fifteen years improving the steam engine,[62] whereas Mr Boyman Boyman spent £6,000 between 1838 and 1850 on an improvement in steam navigation.[63] William Morgan, who patented his improvement in 1829, received £9,400 in advances to complete and carry out his invention.[64] Woodcroft's revolving spiral paddle for propelling ships, on the other hand, cost only £1,400, whilst Stafford spent only a mere £2,000 on improving carriages.[65] In other miscellaneous trades costs were generally lower. Morton's slip for repairing ships cost £700 to perfect and £2,239 to manufacture.[66] Downton's patented water closet cost £500 for experiments and about £1,000 per annum to manufacture, and Charles Derosne's invention for extracting sugar or syrup from cane juice cost as little as £280 to produce.[67] Without further research on the volume of investment in invention it is difficult to say what is typical, but since inventions are so different in nature it seems unlikely that any useful generalisation will emerge to challenge the view that this kind of investment varied considerably, within and between industries.

The sources of investment were also diverse. For industry generally (and this excludes large capital users such as public utilities), Crouzet suggests that there was a hierarchy of investment sources. Internal sources, he claims, came usually from small artisan manufacturers, merchant manufacturers and manufacturers diversifying into other industries. These sources of investment also changed over time, and as technological change raised the threshold of entry capital came usually from within the industry itself. External sources, on the other hand, were important in the initial stages of a firm's development, and here merchant and commercial capital played a much more significant role than landed capital or banks. As firms expanded, though, these sources became less critical (except perhaps in the short run when credit was tight) and entrepreneurs increased their capital by reinvesting profits.[68]

The finance of invention, in contrast, depended largely on external sources, except in the case of multiple inventors and where manufacturers invented on their own account. Multiple inventors who were able to dispose of their inventions frequently used the proceeds to finance further inventions. The income from direct sale, or licensing, was often quite considerable and at the very least provided funds to carry on with. R. Prosser, who took out eleven patents between 1831 and 1852, made profits on a number of his inventions, and reinvested most of the money in the development of others. Some £20,000, for example, was ploughed back into his method of making iron tubes.[69] Bessemer's bronze powder, which was the 'one great success I had so long hoped for', was certainly used to finance his other inventions:

> the large profits derived from it not only furnished me with the means of obtaining all reasonable pleasures and social enjoyments, but, what was even a greater boon in my particular case, they provided the funds demanded by the ceaseless activity of my inventive facilities, without my every having to call in the assistance of the capitalist to help me through the heavy costs of patenting and experimentation on my too numerous inventions.[70]

Other multiple inventors, like Lemuel Wright, undoubtedly reinvested the income derived from their inventions, but apart from inventions which required little variable capital most inventors, at one time or another, sought external support for experiments and development. Few were as fortunate as Bessemer. According to contemporaries, who were unerringly consistent in their guesses, only one to two per cent of inventions were valuable.[71] 'It is doubtlessly true,' Webster observed, 'that a very small proportion of inventions are of practical utility, and that a small proportion of such as are of practical utility remunerate the inventor.'[72] The testimony before the Judicial Committee of the Privy Council amply justifies these claims and, whilst it seems that only about 30% of all patents effectively survived for longer than seven years,[73] even successful patents had a long payback period. Watt's steam engine did not begin to make profits until the early 1790s.[74] Samuel Morton's patent slip did not provide any return for six years, and over the remaining eight he made a profit of only £5,737.[75] Brunel also argued that none of his patents made anything in the first six years,[76] whilst Richard Roberts reckoned that his self-actor, which was his most successful patent, took nine years to make a return.[77] On average a successful patent would yield no profit until the seventh or eighth year.[78] Manufacturers were obviously better placed and could use profits from their business to finance their inventions, but since the mass of users were generally less interested in producing inventive output, apart from minor improvements, most used their resources to support inventors through the formation of partnerships. According to William Fairbairn, 'it *generally* happens, in the purchase of an invention, that the inventor and the capitalist go into partnership; they divide the profits of the invention, if it is a good one, the capitalist taking the whole risk. In the cases where the inventor has no capital, the capitalist looks to him to perfect the invention, as a matter of business to render it profitable.'[79]

Partnership was the typical form of business organisation during the industrial revolution, and it frequently offered opportunities for developing an invention, especially when one partner was a manufacturer intending to use the invention.[80] Such partners would usually cover the cost of patenting, provide all the capital for experiments and development, defend the patent at law and save the inventor from all sorts of pressures.[81] The partnership formed in 1830 between John Ferrabee and Edwin Budding is fairly typical. Here Ferrabee advanced money for the patent, in return for all the profits until the costs of developing Budding's lawnmower were covered, after which profits were shared equally. Budding was what might be called a sleeping partner, and seems to have had little to do with the manufacturing side of the ventures. By 1858 6,000 machines had been produced, which suggests that Budding received a considerable return. In the meantime, he also formed other partnerships for developing his various inventions.[82] Some inventors, though, were more active. Arkwright, who was financed by two rich merchant hosiers, Jedediah Strutt and Samuel Need, was a dynamic force and had an acute business mind.[83] Richard Roberts also played an active part in the partnership he formed with the Sharp brothers, who in the 1820s were the leading cotton-spinning machine makers.[84] Isaac Holden was another inventor who became involved in manufacturing and who relied on the user firm, in this case Samuel Lister's, to provide the capital. Holden became acquainted with Lister in 1846–47, and they registered a joint patent for woolcombing in 1848. In the following year a partnership was formed, with Lister receiving two-thirds of the profit, as well as 5% for the capital which he provided.[85] For Holden, Lister was 'God's agent', and there is very little doubt that a number of inventors were equally relieved to find a partner willing to back their ideas. F. Muntz was certainly happy enough to join with Messrs Greenfell to produce his copper sheathing for ships, and much the same can be said of P. Martineau, who in 1815 formed a partnership with the sugar refiners Beaumont and Wackerbank.[86]

## Investment in patents 161

Existing user firms were not the only source of investment. Some inventors formed partnerships with users, who diversified from closely related trades to set up new ventures. The partnership between Charles Macintosh and Hugh Hornby Birley, the Manchester cotton spinner, is a good example. Machintosh's rubber solution, patented in 1822, was combined with cotton to produce a rainproof fabric, and although there were initial difficulties in persuading the public that the garment was not injurious to health,[87] it sold widely and 'macintosh' became a household word.[88] The mill, however, was designed so that Birley could easily use it for cotton spinning should the rubber partnership fail.[89] Thomas Hancock, who collaborated with Macintosh, especially in defence of their respective rubber patents,[90] also formed a partnership with a Mr Bewley, a manufacturer of stoppers for bottles, who wished to use india-rubber in combination with ground cork and wood dust to produce a more efficient product.[91] In the paper industry, finance for the Fourdrinier paper machine came from the wholesale paper merchants Fourdrinier and Bloxam, who then set up in manufacturing on their own account.[92] Gamble, who brought Robert's invention from France, was entitled to the first £3,675 earned as profits, after which they were divided equally between the partners. Bloxam and Co. were to carry all bad debts and to receive 5% on the gross sales of machinery, insurance and wages.[93] In the flax industry, Marshall developed Matthew Murray's 1793 invention for a carding engine and a wetting sponge for the spinning frame, and on the strength of the patent induced two Shrewsbury woollen merchants, Thomas and Benjamin Benyon, to join him in partnership. They invested £9,000 in the firm in return for a half share of the profits, and over the next seven years the business grew rapidly.[94]

Occasionally, inventors formed partnerships with men who had no obvious connection with manufacturing. One example is George Spencer, who set up as a manufacturing middleman assembling (rather than producing) parts of rubber springs purchased from other manufacturers. His 1852 patent was financed by Richard Kirkman Lane, in return for 'one moiety of all right, title and interest, benefit and advantage' in the invention, and further capital was obtained by forming a partnership with Compton Reade, a Brighton solicitor, who brought in an initial sum of £500 and later another £900. Spencer received 50% of the profits (which included a royalty paid to himself), whilst Lane and Reade shared the remaining half equally. The partnership with Reade was dissolved soon afterwards, and in 1854 Spencer formed another one with F. R. M. Gosset, who within two years invested £6,614 in what turned out to be a very profitable business.[95] How many partnerships relied upon external sources of this kind is unknown, but it seems to have been fairly rare. Investors who had no direct link with manufacturing industry were generally happy to conclude some informal agreement, and this usually involved little more than developing the invention to a stage where it could be sold or licensed. In the early 1840s a Mr Gardiner, from Manchester, assisted Bennet Woodcroft with £1,000 to finance the development of his paddle wheel.[96] Downton borrowed £550 at 5% from an unknown source to develop his water closet,[97] and Goldsworth Gurney sold shares in his patent for a steam carriage to Sir James Viney (who then sold his share to Sir Charles Dance) and to Mr Hanning, Dr M'Kay and Mr Thisselton without formalising the arrangement into a partnership.[98]

Where the inventor was an employee, arrangements were again usually informal. In some cases the inventor would receive a fairly large payment for his ideas. Messrs Sharp gave Hill, head of the loom department, £2,000–£3,000 for his improvement in the carpet loom, and then used their own resources to develop it.[99] In 1849 E. Hayward, a pattern designer with T. and M. Bairstow, worsted manufacturers, assigned his patent to his employers for £100 per annum; after the invention was successfully developed, this was increased to £500.[100] Whitehouse, whose patent for pipe tubes was taken over by his employer, James Russell, was equally fortunate and received an annuity of £300, increased to £500 after the patent was extended in 1838.[101] The engineer J. Kennedy also came to an agreement with one of his employers, but was rather grudging in his praise: 'I am happy to inform you,' he told Isaac Holden,

> that I have unexpectedly been favoured with a satisfactory agreement with ... [Mr Caird] ... who has bound himself to go along with me in securing letters patent for my contrivance for knitting, upon the distinct understanding that he bears [all of the] costs and expenses upon consideration of receiving two thirds value in the patents obtained, while I receive one third as inventor ... [He] may possibly begin to work on his own for the manufacture of our machines soon after he gets out of the present concern — but he is not all suited to look after the getting up of machinery tho' he thinks himself able to conduct a work of the kind. I will have the same right to make our patent as he will, and may do so somewhere else than with him. So I am still at liberty should you require my services and possibly the manufacture of my machines might be beneficially carried on along with yours in the event of our projects proceeding.[102]

Other employee inventors were less fortunate, largely because those skilled workers who were employed for the specific purpose of developing an invention were not entitled to any extra remuneration. In law they were considered as servants, and few had the capital to prove otherwise.[103] From the evidence there is only one instance where an operative was made a partner. In 1822–23 John Mercer, whilst working for Ford Bros. and Co., discovered a bronze colouring by using oxide of manganese. At first the firm disliked the colour, but Mercer persisted and soon made the unpatented invention profitable. In 1825 the firm made him a partner and his subsequent improvements were then all patented.[104] There are likely to have been other cases where employee and employer formed partnerships, but for the most part it would seem that manufacturers preferred to pay operative inventors an annuity.[105]

Although partnership was the typical form of business organisation during the industrial revolution, some inventors formed joint stock companies to raise large amounts of external capital to finance the development of one or more patents. Prior to 1837, however, very few sought the privilege of corporate status and limited liability, and even fewer turned out to be successful business ventures. The papermaking company formed by Mathias Koops, on the basis of three patents taken out in 1800 and 1801 for making paper from straw, hay and thistles, went bankrupt within three years, despite capital of £71,000.[106] The companies formed by J. S. Langton in 1829 for seasoning wood, and by J. H. Kyan in 1836 for curing dry rot, appear to have been more successful,[107] but in the early nineteenth century (and after) manufacturers distrusted the alleged benefits of full incorporation. Private enterprise was considered economically and morally superior;[108] ownership and control went hand in hand, and individuals were responsible for all losses. In fact, of the 910 unlimited companies formed between 1844 and 1856, most were in insurance, mines, public utilities, markets and public halls or shipping, and almost none in manufacturing.[109] After the 1837 Act, which permitted the Board of Trade to confer corporate status, a number of companies were created on the basis of developing one or more patents. By 1852 some twenty-one had been formed.[110] Table 8 provides a list of those which have been identified as specifically formed to develop inventions.

Few details exist to indicate what kind of problems inventors had in raising capital in this way, or how successful the companies were.

*Table 8   Companies working patents, 1837–1852*

| | | |
|---|---|---|
| 1. | London Caoutchouc Co. | 1837 |
| 2. | Patent Dry Gas Meter Co. | 1837 |
| 3. | Colonial Patent Sugar Co. | 1838 |
| 4. | India Steam Ship Co. | 1838 |
| 5. | Ship Propeller Co. | 1839 |
| 6. | Patent White Lead Co. | 1839 |
| 7. | General Filtration and Dry Extract Co. | 1839 |
| 8. | Kollman's Railway Locomotive and Carriage Improvement Co. | 1840 |
| 9. | General Salvage Co. | 1840 |
| 10. | Patent Rolling Compressing Iron Co. | 1841 |
| 11. | Stead's Patent Wooden Paving Co. | 1841 |
| 12. | Tweeddale Patent Drain Tile and Brick Co. | 1841 |
| 13. | Metropolitan Patent Wood Paving Co. | 1842 |
| 14. | Edinburgh Silk Yarn Co. | 1842 |
| 15. | Electrical Telegraph Co. | 1846 |
| 16. | Timber Patent Co. | 1847 |
| 17. | Lowe's Patent Copper Co. | 1847–48 |
| 18. | Patent Galvanised Iron Co. | 1848 |
| 19. | Price's Patent Candle Co. | 1848 |
| 20. | Siever and Westhead Co. | 1849 |
| 21. | Patent Solid Sewage Manure Co. | 1852 |

William Hindmarch claims, not without some prejudice, that few succeeded: 'It is clear indeed that public companies cannot conduct manufacturing or trading business so efficiently as private persons or ordinary co-partnerships and the machinery necessary for the management and regulation of such companies swallows up all the profit made by their trade or manufacture'.[111] The Marquess of Tweeddale's Patent Drain Tile and Brick Company does, however, provide an interesting illustration of the negotiations which took place before a company was created. Oddly enough, although the company was formed to gain access to capital in excess of that which could be got through an ordinary partnership, the shares were ultimately confined to four shareholders.

Tweeddale took out his patent for a machine to make tiles and bricks for draining land and for use in building in 1836. Some trial tiles had been manufactured the previous year, and these were considered by one tile maker as a decided improvement. If they could be produced quickly and in quantity, he claimed, they would soon replace hand-made tiles.[112] Samuel Hollins was sufficiently impressed to offer to introduce the machine once the patent had been obtained.[113] There was less enthusiasm for the bricks made by machine, and over the next two years Tweeddale attempted to rectify the faults. By 1838 a number of tile machines had been made by the engineer Robert Bridges and were working in different parts of the country. The two leading tile makers in mid-Lothian, Dean and Henderson and Reed and Aberlady, had tried the tile machine and managed to cut their costs by 10s per 1,000 tiles, but elsewhere results were less successful, largely because 'the charge of the machines has been committed to persons ignorant of their construction and unaccustomed to the management of machinery'.[114] The production of bricks, on the other hand, had made no progress and in mid-1838 Tweeddale leased the whole tile and brick patent to J. B. Bernie, P. Greathead and G. Stephens to manage the use and licence of the machines on his behalf.[115]

In August 1838 Tweeddale took out another patent to cover the improvements made since 1836, but the brick machine was still proving a problem and could not compete with cheap hand-made bricks.[116] The licensing of the tile machine was very much more successful, and by the end of the year Greathead estimated that the agreements he was negotiating would yield some £6,000 per annum: 'we are able to do much more than I knew of for the production of the patent'.[117] There were a few complaints about the licence fee, which after some deliberation had been set at 3s per 1,000 drain tiles, but by April no money had come in because of the 'unavoidable delay' in negotiating the details, especially in Scotland.[118] The agents remained optimistic about the growing demand for drainage and, as the harvest for 1838–39 was the worst since 1816,[119] they probably expected to persuade landowners and farmers to invest their increased income in improving the land. The late 1830s were also enterprising years for the farming community, and a number were seeking means of reducing unit costs.[120]

In June 1839 Birnie suggested the formation of a company for 'the purpose of carrying on a considerable establishment of our own, furnishing or letting machinery to individuals and advancing funds for works in cases unattended with risk, and where such encouragement would be likely to promote the use of the Patent'. He had been discussing the proposal with Ogle and James Hunt, the 'greatest brick manufacturers in England', for a month or so, and a number of details had been worked out. The initial capital was to be £500,000, with power to raise more if required, and the new company was to act as assignee of Tweeddale's patents. In return, Tweeddale and his three lessees would receive £20,000 in cash, £140,000 in shares and half of all royalties the company received from granting sub-licences. Limited liability was to be guaranteed by an Act of Parliament. The advantage, Birnie observed, 'from this is that by the introduction of unlimited capital the patent will be brought into full effect several years earlier than can well be expected otherwise'.[121] Birnie agreed to act as company secretary, whilst Greathead and Stephens were prepared to be directors.[122]

Tweeddale appears to have been happy with the proposal, and negotiations continued. In August Robert Stephenson was called in by Hunt to assess the machines, and 'after a slight examination declared the principle perfect', but noted that the brick machine required some modification.[123] In the event it was decided to separate the tile and brick machines in 'our dealings with the public' and Hunt agreed to improve the machine, with the aid of Stephenson and Brunel, at his works on the Great Western Railway at Chippenham, in return for some undisclosed 'privilege to which persons in an ordinary scale of business could not pretend'. Greathead urged Tweeddale to give the 'largest encouragement to the proceedings', and tactfully warned that the 'very high mechanical character' of the machine would not be altered in principle. Hunt, he repeatedly stressed, was a skilled engineer and of 'the first class as to respectability and means'.[124]

The separation of the brick and tile machines led to the first alteration in the financial proposals tentatively agreed in June. Now Tweeddale was to get £5,000 in cash, and a further £5,000 when the Act of corporation was passed, plus £30,000 in shares: £40,000 in all for half the patent. At this stage Birnie was not disposed to accept less, and reckoned that shares were more valuable than an equivalent amount in cash; especially as the 'public taste [for] draining is advancing so rapidly' and would make the company 'one of the most profitable undertakings of the day'.[125] In the same month Earl Grey consented to let the proposed company erect a tile works on his estate, instead of building one on his own account, and Birnie claimed a large number of Northumberland landowners were convinced of the advantages of draining their land.[126]

A great deal of this was sales talk, and in October further concessions were made. Hunt would now grant sub-licences only at his discretion and appears to have been ambivalent about developing the brick machine, because bricks made by hand were cheaper and because the trade was reluctant to adopt the new system. He also proposed to confine the shareholders in the company to his brother, George Glyn and Stephenson.[127] Since this made the unlimited capital rather limited, new financial terms were offered in the following month. Instead of the total of £10,000 cash payment Tweeddale was now to receive £5,000 and an additional 6*d* per 1,000 tiles. Birnie, always optimistic, estimated that this would bring in a sum of between £50,000 and £65,000 over the remaining eleven years of the patent's life, and provide ample compensation for the £40,000 Tweeddale would otherwise have got in cash and shares. Birnie insisted that the company would be 'unique (for a long time at least) as it will consist of persons known to and confiding in each other' and would have sufficient capital to market the patent.[128] To persuade Tweeddale that these terms should be accepted, Birnie and Greathead wrote separate letters the next day warning that three patent tile machines were passing through the Patent Office, and that this 'appearance of competition will probably occasion a hurtful delay with the persons with whom we are now dealing. It is, therefore, of great consequence that we shall lose no time in linking ourselves with powerful and influential people before any unfavourable impressions can be created.'[129]

In December, when negotiations were nearly broken off, further concessions were made and there was talk of Tweeddale surrendering £2,000 out of his cash payment to cover the expenses of applying for the Act.[130] In the new year, however, all appeared settled and arrangements were made with Lord Lyndhurst to push the Bill through Parliament. The 'great cry now raised against the Corn Laws' filled Birnie with renewed confidence, and he saw no reason why the Bill should fail.[131] In late January Ogle Hunt began preparing a pamphlet to advertise the company and on 1 February 1840 articles of agreement were drawn up, transferring the patents to the Hunt brothers, Glyn and Stephenson. Tweeddale received a £5,000 cash payment on the 17th, and agreed to accept 2*s* for every 1,000 ft of drain tiles, paving tiles and any other articles, of which a third went to Birnie, Greathead and Stephenson as lessees. All the royalty improvements made by Tweeddale would be shared equally with the company, which had the exclusive power to grant licences at 4*s* per 1,000 tiles, and sub-licence at a premium decided by a referee. The lessees were responsible for supplying three tile machines at cost price and for arranging the construction of twenty-six more, sixteen by Robert Bridges and ten by the London firm, Cottan and Hallam. Tweeddale was permitted to produce tiles for his own use in Scotland, but subject to 1s 6d royalty per 1,000 ft if they were sold. This condition also applied to the brick machine, although here Tweeddale received only two-thirds of a 6d per 1,000 royalty. He and the lessees, were guaranteed a yearly premium equivalent to the manufacture of 20 million bricks of statute measurement. The company was responsible for defending the patents at law and for erecting tile works and, in the case of the brick machine, agreed not to assign the rights to more than five persons.[132] In March Tweeddale wrote to Lord Shaftsbury consenting to the formation of the company, and on the 10th the Bill was given its first reading. On 4 April the Tweeddale Patent Drain Tile and Brick Company became a legal entity.[133]

By August 1841 the company had erected twenty tile works in fourteen counties, and had its own establishment in London.[134] No works were erected in Ireland as deliberate policy, but Twyman, one of Tweeddale's first licensees, was induced to go there to license use of the machines, although at the lower royalty rate of 2s per 1,000.[135] How the company fared thereafter is unknown. Birnie continued to seek the patronage of the 'nobility and other great landowners', and rallied support for Pusey's drainage Bill because as he admitted, there was 'nothing unreasonable in seeking private advantage and benefiting the country at the same time'.[136] The 1840s were, moreover, propitious years for the extension of drainage schemes, and when the government offered financial assistance to appease the landed interest over the repeal of the corn laws in 1846 the manufacture of tiles boomed.[137] In these circumstances there is every reason to assume that the company, being one of the first in the field, proved a profitable concern.

Although the evidence used in this chapter is unsatisfactory, a number of conclusions can be drawn. Firstly, and perhaps most obviously, the cost of producing and developing inventions varied considerably. Secondly, though multiple inventors were able to plough back profits from the sale and licence of their inventions, and inventor/manufacturers were able to use profits from other parts of their business, a great deal of investment in invention came from external sources. User manufacturers would appear to have been the most important

# Investment in patents

source, with partnership the most appropriate method of developing the invention. Thirdly, few were willing to invest in inventions unless protected by patent. Without this security inventors stood virtually no chance of interesting innovators or capitalists and, as Goldsworth Gurney observed, 'without capital I can do nothing'.[138] Finally, there is scant evidence to support Landes's view that inventors had little difficulty in finding funds for development. Convincing investors that they should invest in inventive activity was not easy, except perhaps where an invention was obviously valuable, and even here many users especially were reluctant to break with old methods.

## Notes

1 Ashton, *op. cit.*, 1968, p. 76.
2 F. Crouzet, 'Capital formation in Great Britain during the industrial revolution', in F. Crouzet, ed., *Capital Formation in the Industrial Revolution*, 1972, p. 162.
3 Landes, *op. cit.*, pp. 63–4.
4 Mathias, *op. cit.*, p. 37.
5 S. Pollard, *The Genesis of Modern Management*, 1965, p. 289; and 'Capital accounting in the industrial revolution', *Yorkshire Bulletin of Economic and Social Research*, 1963, pp. 75–91; see also N. McKendrick, 'J. Wedgwood and cost accounting', *Economic History Review*, 1970, pp. 45–67.
6 J. K. Langton, ed., *Memoirs of the Life and Correspondence of Henry Reeve*, 1898. Reeve was appointed in 1837 — much to Lord Brougham's disgust — and held the post for almost sixty years. In all this time there is no reference in the memoirs to the proceedings of the Judicial Committee relating to patent extensions or inventive activity; they are almost entirely concerned with political intrigue.
7 *Select Committee on Patents*, Parl. Papers, XXIX, 1864, p. 451. See also Webster, *Reports*, II, p. 121.
8 Landes, *op. cit.*, pp. 63–4.
9 *Select Committee on Patents*, Parl. Papers, XVIII, 1851, pp. 47–8.
10 *Ibid.*, p. 104.
11 Brougham MSS, [675], W. A. Cochrane to Lord Brougham, 26.3.1849, U.C.L.
12 Bessemer, *op. cit.*, p. 33.
13 *Select Committee on Patents*, Parl. Papers, XVIII, 1851, p. 354.
14 *House of Lords Select Committee on Patents*, 1835, H.L.R.O.
15 Boulton to Lord Dartmouth, 22.2.1775; cf. B. D. Bargar, 'Matthew Boulton and the Birmingham petition of 1775', *William and Mary Quarterly*, 1956, p. 37; E. Roll, *An Early Experiment in Industrial Organisation, being a History of the firm Boulton and Watt, 1775–1805*, 1930, pp. 18–19; Dickinson, *op. cit.*, pp. 85–6.
16 E. Holroyd, *A Practical Treatise on the Law of Patents for Invention* 1830, p. 156; J. P. Norman, *A Treatise on the Law and Practice*

*Relating to Letters Patent for Invention*, 1853, p. 145; Hindmarch, *op. cit.*, p. 66; MacLeod, *op. cit.*, pp. 46–7.
17. Boulton and Watt MSS, J. Donaldson to M. Boulton, 28.11.1787, letter No. 140, Birmingham Assay Office; *J.H.C.*, LXXXII, 1827, pp. 471, 481, 486, 504, 533, 585.
18. *Select Committee on Patents*, Parl. Papers III, 1829, p. 91; Dundonald MSS, Dundonald to Lord Liverpool, undated GD233/107/E/3, S.R.O.
19. Carpmael, *op. cit.*, 1832, p. 41; Hindmarch, *op. cit.*, p. 66.
20. P. L. Cottrell, *Industrial Finance, 1830–1914: the finance and organization of English manufacturing industry*, 1980, pp. 42–4.
21. *Select Committee on the Law of Partnership*, Parl. Papers, XVIII, 1851, p. 155.
22. *Ibid.*, pp. 82–3, 98, 161, 169.
23. *Public Petitions* Ref. 6,294 (App. 521) p. 234. The Belfast Chamber of Commerce claimed that there should be no restrictions on the number of persons with an interest in a patent.
24. Crouzet, *op. cit.*, p. 211.
25. L. S. Pressnell, 'The rate of interest in the eighteenth century', in L.S. Pressnell, ed., *Studies in the Industrial Revolution*, 1960, p. 204. Habakkuk is also doubtful about the influence of interest rates, except perhaps in the most severe periods of war. H. J. Habbakuk, 'The eighteenth century', *Economic History Review*, 1956, pp. 434–6. Since there is no certainty about the outcome of investment, there is very little likelihood of it being responsive to marginal changes in the rate of interest, or any institutional changes in the capital market. See C. Freeman, *The Economics of Industrial Innovation*, 1974, pp. 168, 222, 227.
26. Habakkuk, *op. cit.*, p. 162.
27. *Loc. cit.*
28. Dundonald MSS, Dundonald to Retrie, 10.10.1799, GD/107/E/30, S.R.O.
29. *Select Committee on Patents*, Parl. Papers III, 1829, p. 94.
30. Webster, *Reports*, II, p. 8.
31. *Ibid.*, pp. 113, 197, 563–4; see also Brougham MSS, [1,283], J. L. Drake to Brougham, 24.5.1851; [3,974] J. Lang to Brougham, 21.7.1835, U.C.L.
32. Dircks, *op. cit.*, 1867, p. 77; *Select Committee on Patents*, Parl. Papers XVIII, 1851, p. 49.
33. Barlow, *op. cit.*, p. iii.
34. See, for example, Boulton and Watt MSS, W. Hamilton to Watt, 18.9.1801, 30.5.1804; J. Wilkinson to Watt, 23.1.1805, Box 7, III, B.R.L.
35. *Select Committee on Patents*, Parl. Papers, XVIII, 1851, p. 242; *Select Committee on Henry Greathead's Patent*, Parl. Papers, II, 1810–11, p. 387; *Select Committee on Patents*, Parl. Papers, III, 1829, p. 92.
36. Brief, rather uninformative minutes, bound in three volumes, can be found in the Privy Council Office, Downing Street.
37. Webster, *Reports*, II, p. 121.
38. Hindmarch, *op. cit.*, pp. 144–5.

## Investment in patents

39 Brougham MSS, [20,610], J. A. Murray to Brougham, 26.6.1835, U.C.L.
40 Inventors were fairly happy with the Judicial Committee of the Privy Council and many believed they 'could have no better tribunal'. See *Select Committee on the Design Extension Bill*, Parl. Papers, XVIII, 1851, p. 701; *Select Committee on Patents*, Parl. Papers, XVIII, 1851, p. 242.
41 S. D. Chapman, 'Fixed capital formation in the British cotton industry, 1770–1815', *Economic History Review*, 1970, p. 239.
42 R. Boyson, *The Ashworth Cotton Enterprise*, Oxford, 1970, p. 10.
43 Pollard, *op. cit.*, 1968, pp. 113–15.
44 S. Pollard, 'Fixed capital in the industrial revolution in Britain', *Journal of Economic History*, 1964, pp. 299–314.
45 Cottrell, *op. cit.*, p. 32.
46 S. D. Chapman, 'Financial restraints on the growth of firms in the cotton industry', *Economic History Review*, 1979, p. 69; Cottrell, *op. cit.*, p. 32.
47 *Minutes of the Proceedings of the Committee of the House of Lords*, Parl. Papers, XIV, 1807, p. 12; *J.H.C.*, LXII, 1807, pp. 169–70. *Select Committee on Fourdriniere's Patents*, Parl. Papers, XX, 1837, pp. 37–93, esp. p. 45. See also Brougham MSS, [43,768, 45,344], J. Prince to Brougham, 1.4.1833, 1.8.1833, U.C.L.
48 Coleman, *op. cit.*, 1958, p. 232.
49 Webster, *Reports*, I, pp. 473–8.
50 *Ibid.*, II, pp. 6–18.
51 *Ibid.*, I, pp. 196–9.
52 *Ibid.*, II, p. 564.
53 *J.H.C.*, LIII, 1798, p. 372.
54 W. Pole, ed., *Life of Sir William Fairbairn*, 1970 ed., p. 32.
55 Luckhurst, *op. cit.*, p. 29.
56 *J.H.C.*, XXXVIII, 1782, pp. 882, 897; Fitton and Wadsworth, *op. cit.*, p. 93.
57 *Minutes of Evidence of the House of Lords Select Committee*, Parl. Papers, II, 1801, pp. 137–43; see also *J.H.C.*, XLIX, 1774, p. 347; *J.H.C.*, LVI, 1801, pp. 178, 271–2, 417, 456.
58 Webster, *Reports*, II, pp. 573–5.
59 *Select Committee on Arts and Manufactures*, Parl. Papers I, 1824, p. 309.
60 Webster, *Reports*, I, pp. 486–8, 572.
61 *J.H.C.*, XXV, 1775, pp. 142, 168; J. E. Cule, 'Finance and industry in the eighteenth century: firm of Boulton and Watt', *Economic History*, 1940, p. 320.
62 Brougham MSS, [6,640], Rosser to Brougham, 15.7.1833, U.C.L.
63 PRO/BT/1/4128/50, piece 481.
64 Webster, *Reports*, II, pp. 724–9.
65 *Ibid.*, II, pp. 23–4, 563–4.
66 *Select Committee on Morton's Slip*, Parl. Papers, VI, 1832, pp. 302–3.
67 Webster, *Reports*, II, pp. 1–4, 565–7.
68 Crouzet, *op. cit.*, pp. 167–8, 172, 182–3, 188.
69 *An Appreciation of R. B. Prosser*, MS/608, p. 39, M.R.L.

## 172 Patents and inventive activity

70 Bessemer, *op. cit.*, pp. 59, 81. Even Bessemer's later inventions were financed occasionally with outside help. Chance, for example, assisted him for his experiments in glass. Barker, *op. cit.*, pp. 440–4.
71 *Economist*, 26.7.1851; *Hansard*, CXVIII, 1851, p. 13; *Select Committee on Patents*, Parl. Papers, XVIII, 1851, p. 587; *Select Committee on Patents*, Parl. Papers, XXIX, 1864, p. 116; see also Brougham MSS, [45,492], J. Farey to Brougham, 21.8.1833, and [5,312], J. Ford to Brougham, 14.9.1835, U.C.L.
72 Webster, *op. cit.*, 1853, p. 21.
73 After the 1852 Reform Act patents could be renewed after three and seven years. Between 1852 and 1876 approximately 98% of patents survived for three years, 28–30% for seven years, and only 10% survived the full fourteen years. On the not unreasonable assumption that the rate of survival was similar before 1852, it is clear that the *effective* patent population was much lower than represented by the statistics. *Eighteenth Report of Comptroller General of Patents, Designs and Trade Marks*, Parl. Papers, XXIII, 1901. See appendix B.
74 Cule, *op. cit.*, p. 332.
75 *Select Committee on Morton's Patent Slip*, Parl. Papers, V, 1832, p. 303.
76 *Select Committee on Patents*, Parl. Papers, III, 1829, p. 39.
77 *Select Committee on Patents*, Parl. Papers, XVIII, 1851, pp. 426–7.
78 *Ibid.*, pp. 190, 195, 242, 586.
79 *Ibid.*, p. 414. See also Webster, *op. cit.*, 1853, pp. 26, 31; *Minutes of Evidence taken before the Lords Select Committee on Patents Amendment Bill*, 1835, evidence of J. Farey, H.L.R.O.
80 See above.
81 *Mechanics Magazine*, LIII, 1850, pp. 415–16.
82 Budding MSS, TR RAN/SP3/137, 18.5.1830, Reading University.
83 Fitton and Wadsworth, *op. cit.*, pp. 47, 50, 60, 63–4, 77–8, 81.
84 Musson and Robinson, *op. cit.*, 1969, p. 478.
85 E. M. Sigsworth, 'Sir Isaac Holden, Bt: the first comber in Europe', in Harte and Ponting, *op. cit.*, p. 345; Burnley, *op. cit.*, p. 299.
86 *Beaumont* v. *George*, 1815, Carpmael, *Reports*, I, p. 294.
87 It was widely believed that it stopped perspiration. Some believed that the 'perspiration of the wearer met the rain half way', cf. T. Hancock, *Personal Narrative of the Origin and Process of the Caoutchouc or India-Rubber Manufacture in England*, 1857, p. 54.
88 H. Scherer, 'The mackintosh: the paternity of an invention', *Transactions of the Newcomen Society*, 1952, pp. 77–87.
89 G. Macintosh, *Biographical Memoir: Charles Macintosh*, 1857, p. 82; Woodruff, *op. cit.*, p. 5.
90 According to Scherer there was never a formal partnership agreement between Hancock and Macintosh, *op. cit.*, p. 87.
91 *Hancock* v. *Bunsen, Newtons Journal*, XL, Conjoined Series, 1852, p. 239.
92 Coleman, *op. cit.*, 1958, pp. 163–9, 179–83, 235–6.
93 *Minutes of the Proceedings of the Committee of the House of Lords*, Parl. Papers, XIV, 1807, pp. 25–6.
94 Rimmer, *op. cit.*, p. 44.

## Investment in patents

95 Payne, *op. cit.*, pp. 3–6. Lane also sold his right in the patent to Spencer shortly after the partnership agreement. Spencer bought the rights for £1,300 and was assisted by William Dalton, who then received a half share in the patent.
96 Webster, *Reports*, II, p. 23.
97 *Ibid.*, pp. 565–7.
98 *Select Committee on Mr G. Gurney's Case*, Parl. Papers. XI, 1834, pp. 7, 20, 24, 45.
99 *Select Committee on Papers*, Parl. Papers, XVIII, 1851, p. 347.
100 Bairstow MSS, G. Spense to Bairstows, 4.2.1862,2(159)d, Leeds City Library.
101 Webster, *Reports*, I, pp. 473–8.
102 Holden MSS, J. Kennedy to I. Holden, 5.12.1843, Box VI, Bundle 4, B.L.
103 TR/RAN COJ/13 *Patent Rights of Master and Workmen*, J. Pontifex to R. G. Ranson, 13.12.1839, Reading University. See also *Hill* v. *Thompson*, *Bloxam* v. *Else* and *Minter* v. *Wells*, cf. Godson, *op. cit.*, 1840 ed., pp. 27–8.
104 *Select Committee on Patents*, Parl. Papers, XVIII, 1851, p. 286.
105 *Ibid.*, p. 180.
106 *Select Committee on M. Koop's Patent*, Parl. Papers, III, 1801, pp. 127–33, 41 George III, cap. CXXV; Coleman, *op. cit.*, pp. 172–3. See also W. H. Chaloner, who argues that Koops 'more than any other individual deserves the title of founder of the modern paper industry', *People and Industries*, 1963, p. 103.
107 *L.J.*, LXI, 1829, pp. 494, 500, 522, 536, 545, 555, 559, 588; 10 George IV, cap. CXXV; Earl of Dartmouth MSS, D564/11/3, 5, S.C.R.O.; 6–7 William IV, cap. XXVI.
108 Cottrell, *op. cit.*, p. 41.
109 *Ibid.*, p. 44.
110 This small number suggests that Clapham was mistaken in suggesting that *many* patentees sought charters of incorporation after the repeal of the Bubble Act in 1825, *op. cit.*, pp. 42–3.
111 Hindmarch, *op. cit.*, p. 68.
112 Tweeddale MSS, T. Shutley to R. Brown, 15.11.1835; R. Brown to Tweeddale, 27.11.1835, S.N.L.
113 Tweeddale MSS, S. Holmes to Tweeddale, 26.10.1836, S.N.L.
114 Tweeddale MSS, R. Bridges to Tweeddale, 9.5.1838, S.N.L.
115 Tweeddale MSS, J. B. Birnie to Tweeddale, 21.5.1838, S.N.L.
116 Tweeddale MSS, Birnie to Tweeddale, 5.3.1839, 11.3.1839, S.N.L.
117 Tweeddale MSS, Greathead to Tweeddale, 30.11.1839; Birnie to Tweeddale, 7.12.1838, 18.1.1839, 25.1.1839, S.N.L.
119 Matthews, *op. cit.*, p. 29.
120 Lord Ernle, *English Farming Past and Present*, 1961 ed., pp. 364–7; J. D. Chambers and G. E. Mingay, *The Agricultural Revolution*, 1966, p. 130.
121 Tweeddale MSS, Birnie to Tweeddale, 4.6.1839, S.N.L.
122 Tweeddale MSS, Birnie to Tweeddale, 7.6.1839, S.N.L.
123 Tweeddale MSS, Birnie to Tweeddale, 17.8.1839, S.N.L.

124 Tweeddale MSS, Greathead to Tweeddale, 13.9.1839, S.N.L.
125 Tweeddale MSS, Birnie to Tweeddale, 14.9.1839, S.N.L.
126 Tweeddale MSS, Birnie to Tweeddale, 27.9.1838, S.N.L.
127 Tweeddale MSS, Birnie to Tweeddale, 10.10.1839, 18.10.1839, 26.10.1839, S.N.L.
128 Tweeddale MSS, O. Hunt to Tweeddale, 16.11.1839; Birnie to Tweeddale, 21.11.1839, S.N.L.
129 Tweeddale MSS, Birnie to Tweeddale, 22.11.1839; Greathead to Tweeddale, 22.11.1839, S.N.L.
130 Tweeddale MSS, Birnie to Tweeddale, 4.12.1839; O. Hunt to Tweeddale, 19.12.1839, S.N.L.
131 Tweeddale MSS, Birnie to Tweeddale, 23.1.1840, S.N.L.
132 Tweeddale MSS, Abstracts of Agreement, 1.2.1840, S.N.L.
133 Tweeddale MSS, Tweeddale to Shaftsbury, March 1840, S.N.L.; *L.J.*, LXXII, 1840, pp. 105, 107, 166, 178, 232.
134 Tweeddale MSS, *Tweeddale Patent Drain Tiles*, 1841.
135 Tweeddale MSS, O. Hunt to Tweeddale, Sept. 1841, S.N.L.
136 Tweeddale MSS, Birnie to Tweeddale, 29.1.1841, S.N.L.
137 B. A. Holderness, 'Capital formation in agriculture', in J. P. P. Higgins and S. Pollard, eds., *Aspects of Capital Investment in Great Britain, 1750–1850*, 1971, pp. 174–7.
138 *Select Committee on the Steam Carriage*, Parl. Papers, XIII, 1835, p. 31.

# 9

# Patentees, competition and the law

In their book *The Rise of the Western World*, D. C. North and R. P. Thomas argue that the usual textbook causes of growth — innovation, capital accumulation, economies of sale, foreign trade — are not causes of economic growth. The real cause, they allege, is the institutional setting within which individuals are 'lured by incentives to undertake socially desirable activities'. That is, an efficient economic institution must be devised to 'bring social and private rates of return into closer parity'. The institutional setting they have in mind is one where property rights are clearly defined and reign supreme.[1] Writing two hundred years earlier, Adam Smith argued in a similar vein: 'That security which the laws in Great Britain give to every man that he shall enjoy the fruits of his own labour, is *alone* sufficient to make any country flourish.'[2] When the laws relating to private property are poorly defined, the incentives which spur individuals to promote the social good become weak, and economic growth tends to fall short of its potential. Men will no longer enjoy the fruits of their own labour or, to switch centuries again, there will be an externalities problem.

This analysis of property rights raises some interesting questions regarding the development of patent law, because it was argued earlier that the interpretation of the law by the courts changed, and was seen to have changed, some time in the mid-1830s. On *a priori* grounds, this would suggest that patents were of less value prior to 1830 and that patentees were very likely 'blackmailed'[3] by the users of invention into accepting disadvantageous terms of trade. Likewise, it would suggest that patentees were able to shift the terms of trade in their favour during the late 1830s and 1840s, and that the now more favourable decisions of the courts allowed them to appropriate a greater return for their inventive output. This chapter will attempt to assess this difficult problem. It will examine what effect the changing quality of patent property rights had on the number of inventions which were patented, and what effect it had on the inventors' ability to 'enjoy the fruits of their labour'. Here, other factors affecting the patentees' propensity to use the courts to determine their rights will be examined. The evidence is again patchy and the conclusions are, therefore, fairly speculative. It is important to stress, however, that this chapter is not concerned with the effects of the patent laws on economic growth.

Property rights theory assumes that economic behaviour is determined by market rather than non-market considerations: by the rules of maximisation rather than altruism. Since it has been argued that inventors acted as economic men motivated by profits there is every reason to assume that the uncertain nature of the law would reduce the number of inventions which were patented. The acceleration of patent activity in the 1830s would also appear to support this contention. In the nineteenth century the percentage growth of patenting increased from 21·6% in 1810–19 and 30% in 1820–29 to 67·7% in 1830–39 and 87·7% between 1840 and 1849. The acceleration of patent activity can, however, be explained in several other ways. Firstly, the change in patent activity disguises the fact that the trend of patenting grew almost exponentially throughout the industrial revolution. Secondly, the economy was far more integrated in the 1830s and 1840s and the increase in patenting could therefore be explained by the growing interrelatedness of technology, and by inventors patenting round existing patents. Thirdly, manufacturing, always the sector of the economy most susceptible to technological change, now generated a far greater proportion of national income than before the Napoleonic wars. The changing structure of the economy, which was in part caused by technological change, made further technological change more likely. The declining importance of agriculture, which by 1851 accounted for only 20% of total gross national income, was, in contrast, a sector of the economy with less potential for technological change. Finally, the changing nature of production within the manufacturing sector itself also increased the potential for technological change. Factory production, especially in textiles, brought together the inventions of steam, iron and machinery on an extensive scale, and most of this change came in the 1830s.[4] In many ways the 1830s were a very crucial decade in the development of the British economy, and this is reflected in the increase in patent activity.

The fluctuations in patent activity also indicate that the changing nature of the law had almost no effect on inventors seeking protection for their inventive output. With very few exceptions the number of patents registered fluctuated very closely with general fluctuations in the major cycles, which were characterised, in the expansionary phase, by large increases in long-term investment.[5]

Table 9 sets out the annual data for the major and minor turning points in the fluctuations of English patents from 1750 to 1850. The most significant feature of the data is that fluctuations in patents almost perfectly reflect the fluctuations in economic life, especially in the nineteenth century, when the random element in economic fluctuations, caused mainly by harvests, gave way to a trade cycle with a regular rhythm and duration.[6] Peaks in patents in 1818, 1825, 1828, 1839 and 1845 coincide perfectly with peaks in the trade cycle, whilst the peaks in patents for 1801 and 1830 only precede the peak in the trade cycle by one year. The only significant divergence between the two series occurs towards the end of the Napoleonic wars, when the trade cycle peak in 1810 precedes the peak in patents by three years. Any interpretation of fluctuations in the eighteenth century must be treated rather more cautiously. The economy was then far less integrated and random changes, dictated largely by the weather, were

Table 9 Annual turning-points in patents and the British trade cycle, 1798–1850

| Troughs in patents | Troughs in trade cycle | Troughs in patents | Troughs in trade cycle | Peaks in patents | Peaks in trade cycle | Peaks in patents | Peaks in trade cycle |
|---|---|---|---|---|---|---|---|
|  |  |  |  |  |  | 1801 | 1800–2 |
| 1757 |  | 1809 | 1808 |  |  | 1806 | 1806 |
| 1762 |  | 1814 | 1811 |  |  | 1813 | 1810 |
| 1765 |  | 1817 | 1816 | 1766 |  | 1816 | 1815 |
| 1771 |  | 1822 | 1819 | 1769 |  | 1818 | 1818 |
| 1775 |  | 1826 | 1826 | 1774 |  | 1825 | 1825 |
| 1781 |  | 1829 | 1829 | 1783 |  | 1828 | 1828 |
| 1788 | 1788 | 1832 | 1832 | 1785 |  | 1830 | 1831 |
| 1793 | 1793 | 1837 | 1837 | 1790 |  | 1836 | 1836 |
| 1797 | 1797 | 1842 | 1842 | 1792 | 1792 | 1839 | 1839 |
| 1804 | 1801 | 1848 | 1848 | 1796 | 1796 | 1845 | 1845 |

Source. W. W. Rostow, *The British Economy of the Nineteenth Century*, Oxford, 1948, p. 33; T. S. Ashton, *Economic Fluctuations in England, 1700–1800*, Oxford, 1959.

far more important. The peaks in both series do, however, coincide during the 1790s. The data also coincide very closely in the troughs. From 1826 all troughs in patent activity coincide with troughs in the trade cycle, whilst between 1801 and 1822 patent troughs tend to lag behind the downturn in the trade cycle by an average of two years. The troughs in the 1790s coincided perfectly. Apart from the last fifteen years of the Napoleonic wars, therefore, patents moved very closely with fluctuations in the economy.

The fluctuations in patent activity suggest that the changing interpretation of the law had little or no effect, and that expectations about future trends within the economy generally were the crucial factor. A few inventors may well have been reluctant to patent their inventions because of the decisions made by judges, but this does not appear to have affected the trend in patent activity. Watt junior, for example, was not entirely sure whether William Murdock's improvement of the steam engine was 'contestable on grounds of prior use', but thought it worthwhile 'to run the risk for the sake of the advantage of a patent'.[7] S. Langton was also aware that the courts had a 'strong leaning against' some patents, yet he still patented his invention for seasoning wood because 'I conceive there is every probability of the proposed patent proving lucrative'.[8] Inventors were, typically, risk takers and probably never anticipated that their inventions could end up in the courts. Few in fact did. Barlow, the editor of the *Patent Journal*, summed up the matter thus:

> There is a story told of a celebrated anatomist, who, from deep study of the wonderful organization of the human frame acquired such a morbid sensibility of the extreme delicacy of the physical machine, that he absolutely declined to eat, lest he should disarrange the whole system. A man who will not take a patent from the sole fear that patents sometimes lead to expensive legal proceedings is as wise as the poor surgeon and acts from as good reason.[9]

Although this was written in 1847, when the law was more certain, it is a sensible and accurate assessment. Patents were, after all, the only reliable method of protecting inventive output and, despite the unpredictability of the courts, they were worth having.

What effect the changing nature of the law had on the market value of patents is a far more difficult question to answer. There is virtually no direct evidence and it is, therefore, almost impossible to assess whether inventors were forced before the 1830s to compromise on

## Patentees, competition and the law

disadvantageous terms. The uncertainty of the law, moreover, was only one of a number of factors which determined the value of inventive output. Patentees also had to consider the cost of legal action, the degree of infringement, the strength of opposition and competition and how important they and others thought their inventions were. Inventors had to take, as Nasmyth suggested, a commercial view of their patents, and if the total cost of infringement remained less than the total cost of enforcing protection, then it was better to keep the 'balance in the pocket'.[10]

There is, however, some very circumstantial evidence to support the view that the law prior to 1830 did reduce the market value of inventions generally. Firstly, it was widely agreed that a patent was of very little value until it has had at least one verdict in its favour.[11] No 'patent,' noted Thomas Morton, patentee of a ship repairing slip, 'can be said to be sure until an unsuccessful attempt to infringe it has been made.'[12] This is confirmed by evidence provided in support of James Turner's patent extension in 1791.[13] Before Turner's 1788 court case against Winter, the production of yellow paint was severely reduced because of competitive infringement by London 'artisifers and Chemyists ... who have taken advantage of his specifications'. Before the court case, output was as shown in Table 10. The first obvious point, then, is that the market value of a patent depended upon a successful case at law: if the courts came down in favour of the patentee this provided him with a means of increasing his bargaining power *vis-à-vis* the rest of the trade and other users of the invention.

Table 10   Output of yellow paint, 1783–1790 (cwt)

| | | |
|---|---|---|
| July 1783 | – July 1784 | 39 |
| July 1784 | – July 1785 | 45 |
| July 1785 | – July 1786 | 71 |
| July 1786 | – November 1786 | 18 |
| January 1787 | – August 1787 | 4 |
| January 1778 | – June 1789 | 6 |

*After the verdict*
| | | |
|---|---|---|
| April | – June 1789 | 150 |
| July | – December 1789 | 100 |
| January | – July 1790 | 233 |
| July | – December 1790 | 154 |

The second point is that in the late eighteenth and early nineteenth centuries the uncertainty of the law was recognised by a large number of patentees. Although some were prepared to use the courts even when they were never quite sure of the outcome, it seems likely that few were willing to take the risk. Since the value of a patent in effect depended upon whether it was successful in the courts, it would follow that inventors suffered a reduction in their bargaining power. Watt was certain that this was the case and in a letter to Edmund Cartwright explained his dilemma:

> the trouble and expenses to which we have been put in defending our patent, will effectually operate as a preventative against our taking a share in any other until the present questions have been decided, for we make no scruple in saying that if the doubts started respecting the law of patents in the Court of Common Pleas should prevail, there exists no patent whatever for an improvement and there can exist none which will bear contesting ... However we cannot bring ourselves to believe that doctrines so prejudicial to the spirit of invention and consequently ... [to] the prosperity of the country, can ultimately prevail, and we are so far from thinking of you, that our patent stands on precarious [?] grounds that we have not a doubt of being able to establish it in the end unless the judges are prepared to annihilate patents altogether. We are now fighting the cause of all patentees of Great Britain, for the objections against us are so very general, that every invention, unless it be of a new substance comprehended within their limits: we are only the advanced guard, but should the decision of the C.C.P. be against us it will be high time for the whole body of patentees to consider how they can best avert the impending danger which involve them, as well as us.[14]

This led Watt to believe that 'at present a pirate of an invention has a manifest advantage over a patentee, by availing himself of some failure in the specification even though his piracy be manifest ... [and it] is unjust to rob a man of his property because he has been ignorant of forms'.[15]

The statistics on patent cases seem to reinforce this view, as the proportion of patents contested increased from 1·5% between 1770 and 1799 to 2·8% in the 1840s. The increase is not in any sense remarkable but, as noted earlier, with a more certain law inventors were probably able to sustain the market value of their patents simply by threatening to use the courts.[16] From the mid-1830s the improved nature of the law was widely appreciated, and most inventors (as well as infringers) were rather more certain of the outcome. Inventors at least thought so.

# Patentees, competition and the law

The changing nature of the law was not the only factor which determined whether patentees used the courts. Legal costs were also important. Going to court was an expensive experience, and only valuable patents were contested.[17] According to John Duncan, a solicitor well informed on patent litigation, 'forty-nine out of fifty disputes failed to go into the courts',[18] and the major reason appears to have been the cost of enforcement. Gravenor Henson complained that 'nine times out of ten it is not merit but money which wins',[19] and one writer observed that 'if a patentee is not rich enough to fight he may as well not have a patent at all'.[20]

The actual cost of going to law varied from case to case and depended upon the number of times a particular patent went into the courts. In English law a patent could be cancelled only by a writ of *Scire facias*.[21] In cases of this kind the Crown was the plaintiff and the patentee the defendant. When the patentee was plaintiff, as was usual, a failure in the courts did not repeal the patent. If willing, a patentee was still able to argue his case in the courts.

Before reaching the common law stage, inventors would apply through Chancery for an injunction to restrain infringers. This course of action offered two important advantages. Firstly, it could give patentees instant relief.[22] Secondly, it allowed the Lord Chancellor to appoint qualified inspectors to enter and examine the defendants' accounts and place of manufacture.[23] This provided patentees with valuable information which in normal circumstances could be obtained only by more nefarious means.[24] It also removed the element of surprise, often the major prop in the defendant's case,[25] and gave the patentee time to consider the advantages of taking the matter any further: a breathing space wherein costs and benefits could be balanced out.

Chancery proceedings were expensive and often slow. Charles May of Ransomes and May, the agricultural machine makers, thought the cost of 'getting into Chancery is such that no one would incur it who could avoid it', and he hoped that 'he would not get there again even though his case was as clear a one as possible'.[26] Thomas Morton estimated that it cost him £500,[27] and F. D. Hooper, James Russell's solicitor, calculated that for the six cases in Equity £2,942 was paid out in solicitors' fees alone.[28]

Costs at common law also fluctuated quite considerably, but were generally higher than those in Chancery. J. C. Daniels, for example, paid £750 for his 1827 case, whilst Samuel Clegg had costs of £827 in 1829.[29] James Kay paid £3,700 in his case against the Leeds flax manufacturer John Marshall, and the Birmingham inventor G. F. Muntz had expenses of £10,000.[30] According to one contemporary estimate Watt was alleged to have spent between £30,000 and £40,000,[31] and although recent work has shown the sum involved to have been much lower, estimates range from £5,000 to £10,000.[32] J. Russell, the Wednesbury gas pipe manufacturer, was reputed to have spent half a million pounds over the seven years in which he was engaged in legal battle,[33] but, given that Russell's dispute with Ledsam eventually ended up in the House of Lords, it is quite probable that Russell spent in the region of £70,000.[34] Even this is £30,000 short of the sum Neilson was reported to have spent in the protection of the hot blast.[35]

Going to law also depended on the timing and detection of infringement, and this in turn was affected by the degree of competition between inventors and between inventors and users. There is no adequate measure of competition, but contemporaries frequently stressed the intense spirit of rivalry which existed in the invention market.[36] Since inventions created information which was inexpensive to reproduce, it paid producers and users of inventive output to keep closely in touch with what was happening.[37] James Watt, as usual, grasped the point: 'I hope,' he urged Boulton:

> you will spur on Mr. Handley with the patent and at the same time cause to enquire what new patents are now going through the office, for I do not think that we are safe a day to an end in this enterprising age. One's thoughts seem to be stolen before one speaks. It looks as if Nature had taken up an adversion to monopolies and put the same thing in several people's heads at once ...[38]

J. Hobson Smith, of the Sheffield ironfounders, Stewart Smith and Co., was rather more bitter: 'everything worth pirating is pirated in three months, many things that are very good are pirated in fourteen days after the time of production'.[39] John Martineau knew 'that there was no model or machine in existence in England, of which he could not obtain a model or drawing by paying for it',[40] and John Farey confidently claimed 'that when a new patent is applied for, on any new subject, there are several other applications for patents upon the same subject following each other very quickly ... This is because people keep ears and eyes open.'[41] Gravenor Henson explained that 'there is every reason why I should wish to know when patents are taken out, what is going on at the Patent Office ... [and] to get persons to inspect them'.[42] Thomas Webster observed that he frequently had brought before him five or six patents for the same thing within one or two years, and he knew one case where six patents were taken out for an improvement in railway wheels within six months of each other.[43] In part this intense competition between inventors can be explained by the fact that many were working to solve the same kind of problems: there is very little doubt that simultaneous invention occurred. Yet, as many were aware, cheating was a lucrative trade.

One rather crude index of competition is the number of caveats which were lodged at the Patent Office. This method was extensively used to monitor what was happening in the invention industry, and provided inventors and users with a means of cheating and opposing patent applications before the Great Seal stage.[44] There is no time series of caveats to indicate the number taken out, but since they cost less than five guineas they always exceeded the number of patents sealed. In 1844 — the only year with records extant — there were almost twice as many caveats as patents. Between 1 January and 13 April 229 caveats were lodged, whilst only 140 patents were granted.[45] Many of these were taken out by patent agents to conceal the identity of the petitioners, and a wide variety of industries were covered.

The extent to which caveats were used is reflected in other sources too. In 1771 James Keir entered a caveat for extracting alkali from common salt or sea water,[46] and in 1781 Boulton told Watt that he had:

> consulted Handley about pushing a patent [for the rotative engine] through the office expeditiously ... but he tells me there are two caveats against any patents for steam engines. One is entered by Ewer, the other in the Attorneys name ... we have much at stake and therefore do not let us regard a few hundred pounds to be laid out on patents, and not let Ewer and H[ornblowers] (such blockheads) run away with your lead.[47]

In 1792 Boulton urged Weston to enter a caveat against Hornblower's engine, 'for these things should be cut short in their birth',[48] and seven years later Matthew Murray's patent application for a new method of saving fuel was stopped (temporarily) at the Solicitor General's office by Watt's caveat so that he 'could ferret out the nature of his invention and the manner of its performance'.[49] The use of caveats increased in the nineteenth century as the number of patent agents rose. In 1825 William Carpmael notified Stephen Langton that he had entered a caveat for chains used on railways.[50] In 1839 H. Cooper informed Messrs Sheerman and Evans that Edward Thomas of Swansea was applying for a patent for improvements in the process of making bar iron with anthracite: 'should you consider the above to interfere with either of your caveats of 21.2.1838 and 26.1.1839 an opposition ... must be entered at this office within seven days ... otherwise the patent will proceed'.[51] Eight months later Cooper informed David Mushet that George Godney Dove was applying for a patent for improvements in the manufacture of steel and warned him that this probably interfered with his caveat.[52] In 1848 J. C. Bodmer had Moses Poole enter two caveats, one for steel copper alloy and the other for a process of making steel castings, and all this was done as a matter of course.[53] In short, as James Surrey bitterly observed, 'mercenary adventurers lurking about the premises of the Patent Office' lodged caveats against any invention that was 'likely to be of value'.[54]

Another crude index of competition is the extent to which users of inventive output read and took copies of specifications. Here again, there is no evidence to indicate how often specifications were used, although it was generally alleged that few inventors examined the records of the Patent Office. Carpmael estimated that only one in fifty inventors bothered to see if their inventions were described in earlier specifications.[55] Users, on the other hand, referred to them frequently. 'It is constantly the practice,' Carpmael wrote:

> as soon as a new and useful invention comes out, for persons in the particular branch of the trade to which it relates to get copies of the specification with a view to take the opinions of scientific individuals acquainted with the law of patents to ascertain whether the specification is so drawn as safely to secure the invention; or whether the same might be infringed; with the possibility of setting up a good defence, in case of proceedings being taken by the patentee.[56]

William Fairbairn was also one of many inventors who realised how 'very frequently people are upon the watch to see whether there is any imperfection or any flaw in the clauses of the specification, and they take advantage of it'.[57] Since specifications cost very little to acquire, they provided the cheapest means of establishing (if not always successfully) what inventions were meant to do,[58] and did not alert inventors in the way that, for example, bribing skilled labourers did.

Before the emergence of patent agents, who provided lists of old specifications and warned clients of new ones, inventors and users of inventions would generally employ friends or solicitors to visit the Patent Office. Boulton, for example, suggested to Watt that 'we should employ Handley to look into all the Patent Offices once a month to inform us what is going on in the engine way'.[59] Wedgwood similarly urged his Liverpool partner, Thomas Bentley, to send him a copy of Count de Laucaquais's patent, 'letting the cost be what it will'.[60] In the bobbin net industry specifications of Heathcoat's two patents, taken out in 1808 and 1809, were freely circulated. In 1810 Wittacker, an employee of Messrs Nunn, Brown and Freeman, obtained a copy from a Mr Cocker, and in the following year Brown secured a patent in his own name. William Morely, a machine maker and partner of Messrs Kendall and Allen, was another making use of the specification, and by 1816 — the year Heathcoat moved to Tiverton — no fewer than 156 were infringing his patent.[61] In 1815 Josiah Guest received a copy of Anthony Hill's specification 'which fully explains his process' and returned to William Taitt a copy of Joseph Hately's 1802 patent.[62] John Marshall, who was always on the look-out for new processes, made use (and sometimes successfully) of most specifications relating to the spinning of flax.[63] Joshua Richardson's patent for chains was in use on the Leeds and Selby, and Newcastle and Carlisle railways within two months of his specification being enrolled,[64] and in 1840 James Hunt informed Lord Tweeddale that he had obtained a copy of a specification which he reckoned to infringe his brick and tile patent.[65] In 1864 Robert Macfie told the Select Committee on Patents that he had been buying patent specifications for over twenty years.[66] The growing number of patent and mechanical journals in the early nineteenth century also published specifications (sometimes in an abridged form), and many newspapers, local and national, listed the number of patents granted.[67] The Patent Office, as many recognised, 'was open to the world'.

The specification, therefore, provided competitors with reasonably easy access to the latest inventions, and it seems probable that these were examined *before* resorting to other and more costly means of acquiring information on inventive output. Potential competition is, of course, not synonymous with actual competition, and the degree of infringement depended upon many other factors: the structure of the market and the intensity of competition, the elasticity of substitution of new for old capital and capital for labour, relative factor prices, profit expectations, whether the user industry was expanding or contracting, the probability of an expensive and unsuccessful lawsuit, and the individual skills of the inventor. For patentees, though, the sooner a patent came before the courts the greater the likelihood of protection costs being less than the cost of infringement.[68] Naturally, patentees could not always determine when to go to law. That depended upon the timing of infringement and the speed with which it was detected. But from the evidence it seems that when

## Patents and inventive activity

all other alternatives were closed patentees used the courts at the earliest possible moment.

Since patents protected property rights only for fourteen years the distribution is an expected one, as can be seen in Fig. 1. In the first few years of a patent's life diffusion is likely to be slow. The mass of users were not always aware that inventions were available, and it is likely that the total cost of adopting the new technique exceeded the average cost of existing technology. As a result, patentees were more likely to use the courts because of the small number of infringers and because of the example it would set. So long as patentees could maintain a

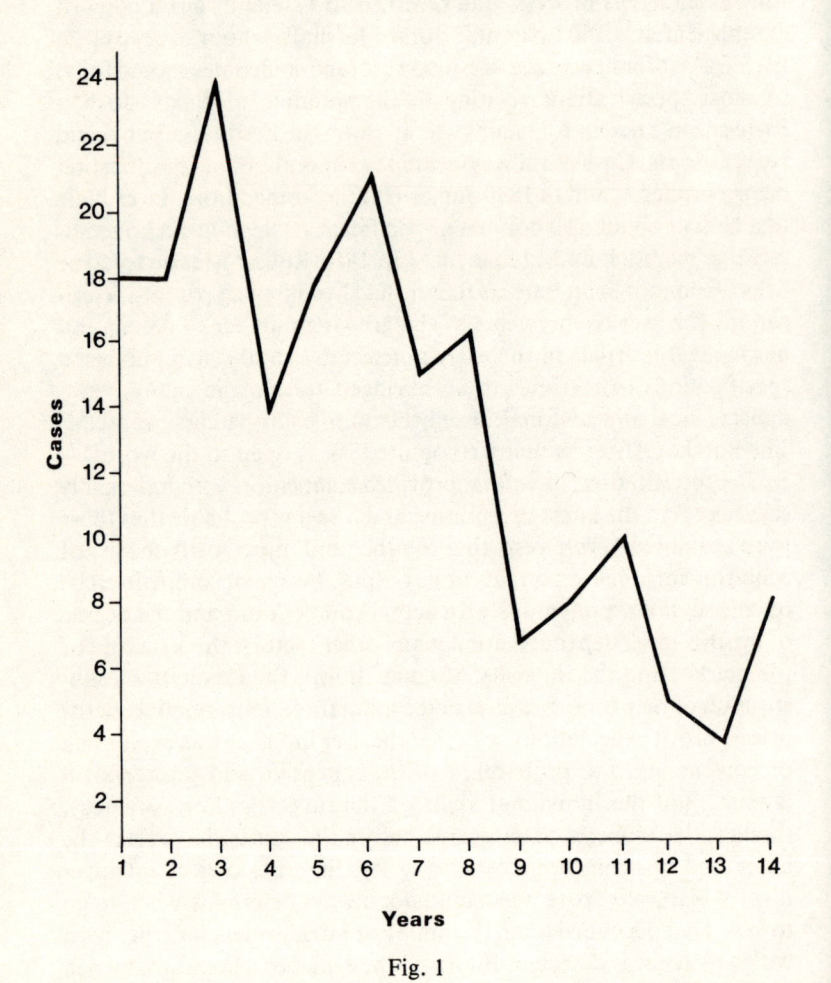

Fig. 1

high rate of detection, the protection of patents at this time was clearly profitable. With the removal of technical teething problems, and with competitors more aware of the existence of new techniques, the rate of diffusion from the fourth to eighth year almost certainly increases. Court cases remained relatively high, but as there were now likely to be larger numbers of infringers, and as the patent had only half its life to run, patentees no doubt tried to negotiate a royalty rate, which would benefit all the parties involved. For the last five years of the patent the number of cases decreased; the cost of protection clearly exceeded the cost of infringement. The slight upturn in the number of patents contested in the final year of the patent's life is difficult to explain, but since the cost of a lawsuit could be used as an argument for having the patent extended, some inventors probably thought it was worth taking the risk of going to law.

Another factor which could affect the value of a patent was the strength of opposition, especially when a large part of the trade united against the inventor. How frequently this occurred is difficult to say. Manufacturers were generally reluctant, on principle, to band together and in practice many such associations soon collapsed. Price associations in the early nineteenth century were difficult to maintain,[69] and even when manufacturers formed Masters' Associations to oppose trade unions and strikes they usually required bonds and fines to penalise those who broke ranks.[70] In ordinary circumstances manufacturers generally practised what they preached: wages, prices and output were determined by supply and demand and best left to the higgling of the market.

The tenets of political economy were, however, fairly malleable. When it suited them, manufacturers could use them to justify almost anything, and in the case of patents they felt free to oppose, with organised resistance, what they considered to be an inventor's monopoly. Manufacturers, John Farey told the 1835 Select Committee, 'wear out the purse of the patentee and almost always by means of a stock purse of the majority of the whole trade, so that the loss to individuals of the opposition is trivial in the aggregate, when to the other parties the expense is so great that it is difficult for patentees to raise the money'.[71] Henry Dircks made the same point some years later:

> manufacturers can ... combine and determine to use an invention in direct opposition to patent monopoly; and, indeed, as they have done in some cases, they can agree to defend each other in any action at law, and even strive to increase the patentee's legal expenses to maintain the alleged

rights. And this may be done with apparent fairness; because they may shew that, considering the patentee exorbitant in his demands, they have offered him the amount at which they themselves assess his claim. It is not pretended, perhaps, that they could not pay his claim, and still realize more than heretofore by their products ... they show that the invention is not costly ... they object to share with the inventor any percentage of the increased value they themselves obtain on their products through his ingenuity [and] they consider [it] as a tax on their own labour.[72]

Collusive action by manufacturers can be identified as early as the 1750s. Jedediah Strutt had been warned by Dr Benson that his inventiveness might give rise to 'innumerable enemies' in the trade. And whilst arranging his court case against Francis James and William Taylor, Strutt sought out a Mr Spackman, a pewterer, 'who has a patent for turning oval dishes and who had some opposition from the rest of the trade but he made 'em pay all the costs and ask his pardon so that now he is more firmly established than if they had not disturbed him'.[73] Dolland's patent for optical instruments was also attacked by the London trade and 'had two tyrals [sic] and got them both'.[74] James Hargreaves, whose jenny was widely used despite his 1770 patent, was less fortunate and declined to defend his rights against a coalition of some seventy or so leading Nottingham hosiers. Although none of these can be identified, 'the association would no doubt have been sufficiently powerful to intimidate all but the most resolute patentee'.[75] Arkwright met with equally hostile opposition and in 1785 lost both his patents.

The facts concerning the dismissal of Arkwright's patents are well known, and the opposition which his 'tyranny' aroused seems to have been justified. There is little doubt that a number of spinners were using Arkwright's model in the late 1770s,[76] and it seems that a number of local merchants raised funds to support Colonel Mordaunt when Arkwright brought an unsuccessful action against him in 1781.[77] This defeat encouraged manufacturers to use the water frame freely, and Daniels estimates that £200,000 was invested in machinery over the next four years.[78] Arkwright, though, continued to threaten users with legal action and in 1783 petitioned Parliament to extend his 1769 patent so that it would expire at the same time as his 1775 patent.[79] The *Manchester Mercury* urged opposition to protest against the 'weight of an unjust monopoly',[80] but did not prevent the renewal of the patent. Flushed with success, Arkwright began negotiating new connections in Scotland[81] and in 1785, when the Manchester men were involved in opposing Pitt's Irish proposals, brought a fresh trial against Peter Nightingale of Nottingham. When the courts decided in Arkwright's favour, alarm spread throughout Manchester. Peel immediately proposed that the trade should set about reversing the decision. He arranged for the twenty-two leading manufacturers to support the expenses of legal action by levying 1s per spindle, and in June 1785 a writ of *Scire facias* demolished both the 1769 and 1775 patents.

Peel's tactics were also employed by John Marshall in 1836, when opposing James Kay's 1825 patent for the process of wet-spinning flax.[82] Here again the trade claimed that Kay was not the real inventor. The process of wet spinning, which was the major invention in the flax industry in the first half of the nineteenth century, led to a boom in the early 1830s and a substantial increase in the number of spindles.[83] According to Marshall, the process — which neutralised a gummy substance called pectose by dipping the flax fibre in cold water prior to the final drawing and twisting - was invented by a Frenchman called De Girard and patented in England by Horace Hall in 1814. Hall sold the patent to a Mr Buck who failed to spin finer than twenty leas and ultimately abandoned the industry.[84] The process then lay dormant until 1825, when Kay patented a modification of the process and added a masticator which permitted the flax to be spun at 2½ in. rather than at 14–24 in. This renewed interest in wet spinning came at a propitious time. The flax industry was badly affected by the recession in the mid-1820s, and failed to recover as other industries did because the falling prices of raw cotton and the wider adoption of the power loom encouraged the substitution of cotton for linen. From the beginning Kay made the invention as public as possible and invited flax spinners to examine the process, but those who took up the invention were unable to make decent fine yarn. Once the early problems were overcome, however, wet spinning spread rapidly and spinners began producing a yarn of 120 leas. In 1831 Kay successfully brought an action against William Renshaw, and it now seemed as if the trade would accept the royalty Kay demanded.[85]

Marshall and Co. were one very important exception and, although the firm suffered quite badly in the late 1820s, they initially refused to take up Kay's invention.[86] In 1827, after a quarrel between John Marshall junior and James Marshall, the firm sought to find its own solution and experimented with hot rather than cold water, which, through a chemical reaction, parted the ultimate fibre, so allowing fine yarn to be spun. How successful they were is difficult to say, but John Marshall senior, who was always willing to try new techniques, was convinced that 'wet spinning is a valuable invention; and we must try and make some profit of it'. Between 1827 and 1836 the firm doubled the number of wet spindles, increasing output by 20% and profits by 60%. By 1835 90% of all spindlage was wet-spinning and the firm was totally reorganised to exploit the increase in demand for fine yarn. Kay was especially successful, and had made significant inroads in Yorkshire and in the Irish and Scottish markets. According to Rimmer this was an important consideration in Marshall's decision to crush Kay by litigation.[87]

In the early part of 1836 Kay sought an injunction against Marshall to prevent further infringement of his patent, but the Lord Chancellor, Cottenhall, decided that the matter should be brought before the common law courts. Marshall now set out to gauge how the rest of the Leeds trade felt, 'for we are fighting their battle as well as our own'.[88] Shortly afterwards articles of agreement were drawn up by spinners using Kay's process, declaring that they 'mutually agreed to assist each other in resisting [Kay's] claim by all lawful means in their power, and to contribute towards the expenses that may be incurred'. A tribute of 2s was levied on each spindle and a bond appears to have been signed to prevent compromise and private contracts with Kay. Fifteen firms signed the agreement: Marshall, 25,364 spindles; Mulholland, 14,644 spindles; J. K. Mulholland, 14,064 spindles; John Boz and Co., 5,358 spindles; James Boomer, 5,000 spindles; Sam Law, 2,800 spindles; E. Grimshaw, 900 spindles, and six anonymous spinners with 18,000 spindles. This was a total of 100,590 spindles, which gave access to £10,059 if required. John Atkinson was appointed executor and administrator.[89] Quite typically, the spinners justified their action in terms of the public good:

> so far from entering into [the agreement] with any view of obtaining an unfair advantage ... they considered that some such combination was necessary in order to meet the Plaintiff on an equality for by making his claim moderate in amount he gave small inducement for any one spinner to oppose it by himself and this though many felt that an unfounded claim ought not to be admitted, he got one by one several spinners to pay him. To meet him with the names of many first rate spinners apparently supporting him and their contributions in his pockets [the] defendants made this defensive agreement with other spinners. They would not have entered upon the opposition of Kay's claim except of Public grounds and with something like general support from others in the same trade.[90]

The trial was held at York on 23 July 1836 and, after a lengthy and thorough examination of the evidence, the judge came down in favour of Kay.[91] This did nothing to change Marshall's mind concerning the validity of the patent, but others were now beginning to question whether continued resistance was worthwhile. B. W. Walker, in a rather ambivalent letter to Marshall shortly before the trial, had already suggested that it might be better to compromise, as Kay's demands were so moderate.[92] This view was also expressed immediately after the trial by the Belfast committee, because it might seem as if the trade was hounding Kay unfairly:

> We are perfectly aware ... of the many difficulties which stand in the way of compromise and reconciliation when proceedings have arrived at such a stage ... Kay stands as yet on very uncertain grounds but still it must be admitted that he has been a useful man to the trade and even if defeated eventually it will be generally received opinion that he was overpowered by the weight opposed to him.

They were willing to agree with Kay if his demands were reasonable and proposed a royalty of 1$s$ per spindle, as against the 6$s$ per spindle Kay demanded, to prevent further litigation.[93] Marshall's agreed it was desirable to end the matter one way or another, but they themselves were not prepared to pay anything for a claim which they considered unjust. They realised that Kay's success would now make him difficult to 'treat with', but were not ready to submit until they had sought the advice of council.[94] 'Our ... motives all along have been to resist what is considered an unjust claim ... not with a view of any penuciary saving to ourselves by incurring a troublesome and expensive lawsuit, but in discharge of what is considered a duty to ourselves and to the trade generally to come forward and assist an unjust imposition.'[95] Mulholland's informed them that Kay was not willing to compromise, and Messrs Ley and Mason wrote on the same matter[96] but, whilst Marshall's were at first prepared to go along with the majority, by November they had reverted to emphasising that it was their duty to oppose for the trade generally: 'having commenced with a full view of the sacrifices of time, trouble and money that would be necessary; these considerations will not weigh with us as inducements to desist ... but if the other parties to the bond [wish] to act for themselves we [are] ready to counsel it'.[97]

What happened thereafter is unclear. In his evidence before the Judicial Committee of the Privy Council in 1839,[98] when his patent was extended for three years, Kay claimed that he received only £500 in licence money between 1837 and 1839 and that only one licence had been issued in 1838. He also claimed to charge only 1$s$ per spindle. This suggests two interpretations. Firstly, the reduction in the licence fee indicates that he compromised with the trade (except Marshall's), and that this compromise took place in 1837, since he had received only £500 in royalties in the two years prior to his application for an extension of the patent. Secondly, it could also mean that he was willing to reduce the cost of licensing to the level recommended by the Belfast committee, but that the trade refused to accept that he was the inventor and pay the revised royalty. Rimmer, who does not deal with the issue at any length, is rather ambivalent, but seems to suggest that the trade supported Marshall (perhaps reluctantly) over the next five years. 'Masquerading as saviours of the industry,'[99] Marshall's were probably able, through a form of moral blackmail, to persuade most spinners to keep the agreement made in 1836. Since the joint legal costs of £2,531 19$s$ 6$d$[100] were only £1,200 less than Kay incurred over a period of five years' litigation (1836–41), the evidence seems to support this second interpretation.[101] In the event, Kay's patent came before the Chancery courts in 1838 and 1839 and was ultimately set aside in 1841 by the House of Lords. Marshall's persistence, together with the weight of the trade, brought down an invention which made the boom in the early 1830s possible. None disputed the importance of the invention and some, like Hives, Marshall's former employee and partner, thought that Kay had made an original contribution.[102] Kay's defeat, though, meant that Marshall's (and others) could pocket the £7,600 they would have had to pay in licence fees.[103]

Evidence for other cases of collusion is less detailed. In 1829 John Lewis, a cloth manufacturer at Brimscombe, near Stroud, was opposed by the rest of the trade when he brought two cases, one against William Davies and the other against N. S. Marling, for infringing his 1818 patent for shearing cloth from list to list by a rotary cutter.[104] Several patents had been taken out for this process since 1811, but by the late 1820s and early 1830s the shearing machines were generally known as 'lewises'.[105] By 1829 upwards of a thousand machines had been sold, and although Lewis 'appears rather to have been in the position of Arkwright, whose main service was the making of previous ideas useful',[106] the court decided in his favour. Fresh evidence against Lewis was discovered after the trial, but an application to try the case again was rejected, probably because the Attorney General (Scarlet) thought that 'a patentee ought not to be driven by the trade to bring several actions in support of his right'.[107] In the lace industry Fisher, who had bought Williams Sneath's patent for improvements in the bobbin net machine, was also persistently infringed against in the early 1840s. Despite his securing an injunction in 1843 the trade formed a combination financed by a general subscription.[108] In 1851 J. Scott Russell claimed he knew of a case in the woollen industry where the trade combined to litigate and eventually forced the patentees into accepting much easier terms.[109] Earlier, in 1843, he also acted as arbitrator in the litigation concerning Neilson's hot blast patent, where again the ironmasters, led by William Baird, combined to remove the 1s per ton royalty which Neilson was charging.[110] Barlow, the patent agent, also lamented the cases of Joseph Heath, whose 1839 patent for improvements in the conversion of iron into cut steel 'was cruelly oppressed by combinations of manufacturers [who] finally ruined him by expensive litigation'.[111] Similar examples can be found in agriculture, although the incidence with which collusion took place seems to have been far less frequent. In the late 1790s, however, Meikle was forced to drop his suit against John Raistrick, largely because several associations formed in Northumberland promised to support financially the defendant's claim that he was the real inventor of the threshing machine. Meikle had also managed to antagonise many farmers and millwrights with threatening letters and thus weakened his position even more. In the event Raistrick bore nearly all the cost, as he was deserted by the farmers who pledged to support the subscription.[112] Bovill was rather more persistent than Meikle, and although a large association of millers sought to bring down his 1846 patent for grinding corn they were 'utterly defeated', despite an outlay of some £40,000.[113]

Collusion was not all one-sided, and patentees would themselves occasionally employ the same tactic. In 1829 John Farey claimed, with some exaggeration, that 'the few patents that have been supported have been commonly sustained by collusion with infringers themselves, [and] after one trial has decided that the patent is not absolutely bad, they combine with the patentee to allow them free use of the patent on moderate terms, and then by making a common purse they prosecute and suppress all new infringements'.[114] In all probability Farey based this assertion on the lace industry, and on the method used by Heathcoat to overcome the problems of the over-expansion of the industry's output. During the period 1810–13 the falling price of lace prompted Heathcoat to take legal action against a number of infringers of his 1809 patent. Successful injunctions were brought against Kendall and Allen and Nunn Brown and Freeman of Warwick, who then became licensees. In 1816 Heathcoat established his right when, in the case of Bovill v. Moore, the judge concluded that both machines in question were an infringement of his patent. Further injunctions were granted against Grace, Standford and Berridge, after which the registration of licensees, as we have seen, increased rapidly. By 1819 the expansion of output exceeded demand, and Heathcoat was finding it difficult to get licence money paid in. A general meeting of patentee and licensees was then arranged and they agreed to establish a 'mart' to control licences and output. A levy was imposed on members, and this was to be used as a means of preventing further infringement.[115] According to Farey — who was well acquainted with the industry — this cartel attempted to limit the use of the invention after the expiration of the patent in 1823, by paying Heathcoat a reduced royalty and having him threaten new entrants with prosecution under a second patent. In the event 'they all laughed at the threat',[116] and between 1823 and 1826 the industry went through an unprecedented boom as 'twist net fever' induced a most unlikely number of bankers, lawyers, physicians, clergymen and landowners to take to the trade.[117]

Unfortunately, Farey does not mention any other cases of this kind, although he claims to have been told that Arkwright attempted to do the same. Most inventors probably only used one or two infringers, with whom they had compromised, to defend the patent. In the early 1770s Wedgwood openly despised 'sniveling copyists' but was prepared to compromise with Neale and Palmer, because they might support him 'in any future tryal',[118] and this appears to have been as close as most inventors came to forming offensive cartels. Where collusion was impossible inventors relied upon threats. A cautionary advertisement in one Birmingham newspaper is perhaps typical:

> Whereas we, the under-named Joseph Rogers and George Sanderson, have been guilty of an infringements [sic] of Mr. Stedmans Patent Knee Buckle, for which he commenced an action against us in his Majesty's Court of Common Pleas; but upon our agreeing to make him a small pecuniary satisfaction, and by paying all the costs of the action hitherto incurred, and on further agreeing to insert this Advertisement as a caution to all persons from being guilty of the like invasion of this patent, he hath agreed to discontinue his said action.[119]

In 1851 Henry Cole told the Select Committee, 'I do not believe that anybody is so fond of litigation as to go into it for its own sake; where you have litigation you have wrongs; the easiest way of getting rid of wrongs is to have litigation; not to be afraid of it.'[120] This is a rather simplistic view. Going to law was a much more complicated affair, and created all sorts of problems for inventors. Most would avoid it if they could, especially if, as was usually the case, there was scope for compromise. 'Few cases,' one eminent patent lawyer wrote, 'require so much care and industry in preparing for a trial as patent actions, in which very nice points of law and difficult questions of fact must be decided between parties; and it will frequently happen that a party will succeed or fail in obtaining a verdict according to the industry with which he has got up his case for trial.'[121] In so far as their value ultimately depended upon the law, patents were generally considered fair plunder. Many users would infringe patents until the inventor actually used or threatened to use the courts, and whether patentees used the law depended, in turn, upon a number of factors. Inventors had to take a calculated risk, based on a crude cost−benefit analysis.[122]

The purpose of this chapter has been to assess whether the changing nature of the quality of patent property affected the market value of patents. Admittedly, the evidence is not altogether convincing, but it seems likely that the uncertain and unpredictable interpretation of the law reduced the bargaining power of patentees prior to 1830 and forced them to compromise at rather unfavourable terms of trade. Whether this was the major determinant of the value of patents is unclear, since it seems that the cost of litigation, the strength of opposition, and the nature of the invention were equally important. Although actions were rare in relation to the total number of patents, this 'should not be taken as an indication that patentees seldom attempted to exploit their rights or that patents were seldom resisted by third parties'.[123] Competition in the invention market was fairly intense and the *unofficial* testing of patents outside the courts was a continuous process, the outcome of which depended upon the relative bargaining strength of the parties involved.

## Notes

1 D. C. North and R. P. Thomas, *The Rise of the Western World: a New Economic History*, Cambridge, 1976, pp. 2−3.
2 Smith, *op. cit.*, I, p. 540. Emphasis added.
3 For a useful theoretical discussion see E. J. Mishan, 'Pareto optimality and the law', *Oxford Economic Papers*, 1967, pp. 255−87.
4 A. E. Musson, 'Industrial motive power in the United Kingdom', *Economic History Review*, 1976, pp. 415−39; V. A. C. Gattrell, 'Labour, power and the size of firms in Lancashire cotton in the second quarter of the nineteenth century', *Economic History Review*, 1977, pp. 95−139; P. Deane and H. J. Habakkuk, 'The take-off in Britain', in W. W. Rostow, ed., *The Economics of Take-off into Sustained Growth*, 1963, pp. 63−82.

5 Gayer, Rostow and Schwartz, *op. cit.*, pp. ix–x. See also T. S. Ashton, 'Some statistics of the industrial revolution in Britain', *Manchester School*, 1948, pp. 214–34.
6 For a general discussion see D. H. Aldcroft and P. Fearon, eds., *British Economic Fluctuations, 1790–1939*, 1972.
7 Boulton and Watt MSS, Watt junior to W. Lawson, 4.4.1799, Parcel E, B.R.L.
8 Langton MSS, Langton to J. Turnley and Sons, 17.12.1827, L.R.O.
9 *Patent Journal*, III, 1847, p. 272.
10 *Select Committee on Patents*, Parl. Papers, X, 1872, p. 181.
11 *Select Committee on Patents*, Parl. Papers, III, 1829, p. 64, *Select Committee on Patents*, Parl. Papers, XVIII, 1851, p. 233; Dircks, *op. cit.*, 1867, p. 95.
12 *L.J.*, LVIII, 1826, pp. 375–6.
13 *J.H.C.* XLVI, 1791, pp. 223, 295–6, 385; *J.H.C.*, XLVII, 1792, pp. 440–1.
14 Boulton and Watt MSS, Watt to Cartwright, 17.3.1795, Letter Book B.R.L.
15 Boulton and Watt MSS, Observations on Miscellaneous Specifications, Box 41; see also J. Watt junior to Rev. J. Gutsch, 5.6.1795, Letter Book; J. Watt junior to W. Lawson, 26.1.1799, Parcel E, B.R.L. See also Brougham MSS, [8,749], W. Ivory to Brougham, 9.6.1835, U.C.L.
16 *Select Committee on Patents*, Parl. Papers, XVIII, 1851, p. 414.
17 *Cornish v. Keene*, Carpmael, *Reports*, I, p. 317; *Select Committee on Patents*, Parl. Papers XVIII, 1851, pp. 266, 280.
18 *Select Committee on Patents*, Parl. Papers, XVIII, 1851, p. 127.
19 Add. MSS. 27,807, Henson to Place, 31.5.1825.
20 *Journal of the Society of Arts*, I, 1853, p. 527.
21 A writ of *Scire facias* repealed a patent in three ways. (1) Where the Crown granted one and the same grant to several persons. Here the first patentee had the right to repeal the others. (2) Where the Crown awarded a grant upon a false suggestion. (3) Where the patent had been granted for an invention which by the law of the land could not be granted.
22 Godson, *op. cit.*, 1840 ed., p. 246. Injunctions were granted only if the patent had been in existence for some reasonable time, 'not of yesterday', and when it was thought the specification was good. *Hill v. Thompson*.
23 Russell MSS, *Russell v. Cowley and Dixon* 1834. Brunel and Donkin were the two inspectors chosen by the plaintiff. Bramah and Clegg were the two chosen by Cowley. All four inspected the respective factories and the machinery in dispute. They were then to place their findings before the Lord Chancellor. Bundle No. 16, 67/16/33, William Salt Library; Hindmarch, *op. cit.*, pp. 296–7.
24 Boulton and Watt MSS, J. Watt to W. Lawson. Writing about an infringement of the Salford Twist Co., Watt notes, 'it will be proper for you to make an attempt to see the engine yourself, and if you do not succeed, employ spies for the purpose or try through the medium of Perrins or

any other workmen to get some of their own people to turn informers', 3.3.1796, Parcel E, B.R.L. Pirated machines were generally carefully guarded. See, for example, W. B. Crump, ed., *Leeds Woollen Industry, 1780–1820*, Leeds, 1931, pp. 188–9.
25 See *Trial of a Cause, op. cit.* After 1835, with the change in the rules of pleading (5 and 6 William IV, C83, s. 5), defendants had to give prior notice of their objections to the patent, and this removed the surprise element.
26 *Select Committee on Patents*, Parl. Papers, XVIII, 1851, p. 369.
27 *Select Committee on Mr. Morton's Patent Slip*, Parl. Papers, XXIV, 1833, p. 300.
28 Webster, *Reports*, I pp. 473–8.
29 Carpmael, *Reports*, I, pp. 453–7; *Repertory of Arts*, V, Third Series, p. 57; *Select Committee on Patents*, Parl. Papers, III, 1829, p. 149.
30 Webster, *Reports* II, p. 111; *Times*, 26.7.1856.
31 *Select Committee on Mr. Gurney's Steam Carriage*, Parl. Papers, XIII, 1835, p. 15.
32 Musson and Robinson, *op. cit.*, 1969, p. 418. Here they estimate legal costs at £5,000–£6,000. In Robinson, *op. cit.*, 1971, the estimate is £10,000.
33 J. F. Ede, *History of Wednesbury*, 1962.
34 F. W. Hackwood, *Wednesbury Workshop*, Wednesbury 1889, p. 97.
35 Macfie, *op. cit.*, 1869, p. 34; *Select Committee on Patents*, Parl. Papers, XXIX, 1864, p. 394.
36 Since patents were granted for a period of fourteen years, the extent of competition between patentees and users of patented inventions depended upon the total number of patents in existence at any one time. See Appendix C for the potential number of English patents in force.
37 See, for example, Dundonald MSS, W. Hamilton to Lord Dundonald, 4.5.1801, GD 233/110/J/I.34, S.R.O.
38 Watt to Boulton, 13–16.2.1782, cf. Musson and Robinson, *op. cit.*, 1969, p. 96.
39 *Select Committee on Arts and Manufactures*, Parl. Papers V, 1835, p. 389.
40 *Select Committee on the Exportation of Tools and Machinery*, Parl. Papers, I, 1824, p. 7.
41 *Select Committee on Patents*, Parl. Papers, III, 1829, p. 18.
42 *Select Committee on Postage*, Parl. Papers, XX, 1837–38, p. 219.
43 Cf. Macfie, *op. cit.*, 1869, p. 35.
44 John Taylor informed the 1829 Select Committee that opposition seldom had the effect of preventing the patent being granted. *Select Committee on Patents*, Parl. Papers, III, 1829, p. 5.
45 P.R.O. HO/LO/I/I.
46 *J.H.C.*, XXVII, 1780, pp. 913–15. In December 1771 Keir wrote to Watt thus: 'When I was last in London I heard it mentioned by some persons curious in chemistry and who especially had made many attempts to discover the method of obtaining the alkali or common salt, that Dr. Black had actually discovered the method. I was afraid lest some persons might prevent our obtaining a patent, if we should think it necessary, by entering a caveat in general against all patents for

obtaining alkali from sea-salt, especially as I found upon Inquiry that you and Dr. Black had not taken that precaution. I accordingly entered a caveat in my own name which caveat shall not prevent you or your friends connected in The Scheme from taking out a patent when you and they think proper.' Cf. Musson and Robinson, *op. cit.*, 1969, p. 359.

47 Boulton and Watt MSS, Boulton to Watt, 1.9.1781, Parcel D, B.R.L.
48 Boulton and Watt MSS, Boulton to Watt, 9.10.1792, 10.10.1792, Parcel D, B.R.L.
49 Boulton and Watt MSS, Watt junior to W. Lawson, 27.3.1799, B.R.L.
50 Langton MSS, Carpmael to Langton, 1.11.1825, L.R.O., Brougham MSS, [21,181], J. Richardson to Brougham, 8.6.1835, U.C.L.
51 Mushet MSS, H. Cooper to Sheerman and Evans, 2.2.1839, D2646/144, G.C.R.O.
52 Mushet MSS, H. Cooper to D. Mushet, 5.10.1839, D2646/147, G.C.R.O.
53 Henderson, *op. cit.*, p. 41; see also Marriner MSS, H. Cooper to Marriner, 15.5.1838, Box 95/1, B.L.
54 Brougham MSS, [8,758], Surrey to Brougham, 12.7.1835, U.C.L.; *Select Committee on Patents*, Parl. Papers, XVIII, 1851, pp. 183–4; *Economist*, 26.7.1851; *Select Committee on Patents*, Parl. Papers, III, 1829, pp. 5, 10, 21.
55 *Select Committee on Patents*, Parl. Papers, XVIII, 1851, p. 42. See also *Select Committee on Patents*, Parl. Papers, III, 1829, p. 103. See also Dundonald MSS, Lord Dundonald to Mr Wilkinson, 17.11.1799, GD 233/109/7/17, S.R.O.
56 Carpmael, *op. cit.*, 1832, p. 45. See also *J.H.C.*, LXXV, 1820, p. 310.
57 *Select Committee on Patents*, Parl. Papers, XVIII, 1851, p. 414; Brougham MSS, [8,764], A. J. Forsyth to Brougham, Aug. 1835, U.C.L.
58 Specifications were enrolled in three offices: the Petty Bag Office, the Rolls Chapel Office and the Enrolment Office. A copy of a specification cost 1s in the Enrolment Office and 3s 6d in the other two. Without the name of the patentee or the date of the patent it was often difficult to find the specification, because these offices held a large class of records, and they were not always indexed. One writer in *The Times* complained that searching for a specification was like hunting for a rat, 'bar the fun', and others complained that the search was unnecessarily tedious. See *The Times*, 2.2.1853; Brougham MSS, [30,824], A. Gordon to Brougham, 19.12.1831, U.C.L.; Gomme, *op. cit.*, 1951–53, p. 163; Collier, *op. cit.*, 1803, p. 19. It is important to note that only Patent Office officials were permitted to make copies of the specification. In 1849 Lord Langdale (Master of the Rolls) opened the Rolls Chapel Office to the public, after which specifications were almost entirely enrolled in the other two offices, so that Patent Office officials could still charge their fee. *Select Committee on Patents*, Parl. Papers, XVIII, 1851, p. 3.
59 Boulton and Watt MSS, Boulton to Watt, 20.6.1781, Parcel D, B.R.L.
60 Wedgwood MSS, J. Wedgwood to T. Bentley, 12.12.1767, John Rylands University Library of Manchester.

## Patentees, competition and the law

61 Felkin, *op. cit.*, pp. 193–226, 245.
62 M. Elsas, ed., *Iron in the Making: Dowlais Iron Co. Letters, 1782–1860* 1960, p. 187.
63 W. G. Rimmer, *Marshall's of Leeds: Flax-Spinners, 1788–1886*, 1960, esp. ch. IV.
64 Brougham MSS, [21,181], Richardson to Brougham, 8.6.1835, U.C.L.
65 Tweeddale MSS, J. Hunt to Lord Tweeddale, 8.10.1840, S.N.L.
66 *Select Committee on Patents*, Parl. Papers, XXIX, 1864, p. 469.
67 The *Repertory of Arts*, first published in 1794 and edited by W. H. Wyatt, was the first to print full and abridged versions of recent specifications. Although not all specifications were published, according to Lord Ellenborough the *Repertory* was a 'mischievous work because it conveys knowledge of English Inventions abroad'. W. H. Wyatt, *A Compendium of the Law of Patents for Invention*, 1826, p. 4. See also *Select Committee on the Exportation of Tools and Machinery*, Parl. Papers, I, 1824, pp. 146–7.
68 Wedgwood MSS, J. Wedgwood to T. Bentley, 13.10.1770, 'I expected no less than what you have wrote me respecting the invasion of our patent, and I apprehend they will persist in it to the utmost so that a trial seems inevitable and if so the sooner the better.' See also Wedgwood to Bentley, 22.10.1770, John Rylands University Library of Manchester; E. Meteyard, *Life and Works of Wedgwood*, 1865, II, pp. 197–8; Tweeddale MSS, J. Hunt to Lord Tweeddale, 8.10.1840, 22.5.1841, S.N.L.
69 Jones, *op. cit.*, pp. 237–53.
70 See, for example, H. I. Dutton and J. E. King, *Ten per Cent and No Surrender: The Preston Strike, 1853–54*, Cambridge, 1981.
71 *The Lords Select Committee on Patents Amendment Bill*, 1835, H.L.R.O.
72 Dircks, *op. cit.*, 1869, p. 16; see also Carpmael, *op. cit.*, 1832, pp. 54, 82–3.
73 Fitton and Wadsworth, *op. cit.*, pp. 35, 42.
74 Strutt MSS, J. Strutt to Wife, 21.6.1766, Derby Reference Library; PRO/PC/1/7/94; MacLeod, *op. cit.*, pp. 100–1.
75 Chapman, *op. cit.*, 1967, pp. 47, 52; and 'The transition of the factory system in the Midlands cotton-spinning industry', *Economic History Review*, 1965, pp. 536–43.
76 *Ibid.*, p. 73.
77 L. S. Marshall, *Development of Public Opinion in Manchester, 1780–1820*, Syracuse, 1946, pp. 179–80.
78 G. W. Daniels, *The Early English Cotton Industry*, Manchester, 1920, p. 102.
79 Pares MSS, T. Pares to J. Pares, 10.2.1783, Derby Reference Library.
80 *Manchester Mercury*, 25.2.1783.
81 Fitton and Wadsworth, *op. cit.*, pp. 85–7.
82 Houghton, Craven MSS, Kay's patent was extended to Ireland and Scotland, 21.5.1825, DDH/574, Lancashire County Record Office.
83 Rimmer, *op. cit.*, pp. 162–228. For a contemporary discussion of technology and changes in flax spinning see A. J. Warden, *The Linen Trade*, 1864, pp. 683–700.

84 Marshall MSS, Memo of Mr Dawson and Mr Younger, E/16/26; MS Note of Individual Witnesses, 5.7.1836, E/16/6; E/16/9. Notes on Kay's Patent, E/16/15; E/15/27, B.L.
85 Marshall MSS, W. Renshaw to J. Marshall, 29.7.1831, E/16/17, B.L.
86 Marshall MSS, J. Marshall junior to J. Marshall. Marshall junior was informed by Solomon Robinson that Kay intended prosecuting infringers in 1827 but 'he will find his work set', 11.11.1827, E/17/17 B.L.
87 Rimmer, *op. cit.*, p. 192.
88 Marshall MSS, J. Marshall to J. Marshall, 9.5.1836, E/17/21, B.L.
89 Marshall MSS, E/16/16, Articles of Agreement, B.L.
90 Marshall MSS, Agreement between Spinners E/16/16, B.L.
91 *Kay v. Marshall*, Webster, *Reports*, I, pp. 75, 213, 396, 401, 409; Marshall MSS, J. G. Marshall to Henry, 23.10.1836, E/17/28, B.L.
92 Marshall MSS, B. W. Walker to Marshall, 20.7.1836, E/16/15, B.L.
93 Marshall MSS, James Campbell to Marshall, 10.8.1836, E/16/15, see also the copy in E/16/14, B.L.
94 Marshall MSS, J. Morall to J. Marshall, 17.8.1836, E/16/14. A copy was also sent to A. Mulholland, B.L.
95 Marshall MSS, J. G. Marshall to J. Marshall junior, 13.8.1836, E/16/14, B.L.
96 Marshall MSS, Ley and Mason to Marshall, 23.11.1836, E/16/4, B.L.
97 Marshall MSS, 23.11.1836, E/16/4, B.L.
98 Marshall MSS, E/17/38, B.L.
99 Rimmer, *op. cit.*, p. 220.
100 These costs include £512 10s 6d for opposing Kay's patent extension. This too suggests that in 1839 the trade was still supporting Marshall's. See E/16/7, B.L.
101 Webster, *Reports*, I, pp. 568–72.
102 Marshall MSS, J. Hives to J. G. Marshall, 16.7.1836. Hives, in fact, refused to give evidence for Marshall in 1836, E/16/19, B.L.
103 This figure is based on the number of spindles Marshall was working in 1836, and at a royalty of 6s per spindle.
104 *Lewis v. Davies*; *Lewis v. Marling*, 1829, Webster, *Reports*, I, p. 492.
105 Julia de Mann, *The Cloth Industry in the West of England from 1640 to 1880*, Oxford, 1971, pp. 303–5.
106 *Ibid.*, p. 303.
107 *Lewis v. Marling*, 1829.
108 *Fisher v. Oliver and Atkins*, 1847; *Newtons Journal*, XXX, 1847, Conjoined Series, pp. 58–9.
109 *Select Committee on Patents*, Parl. Papers, XVIII, 1851, p. 147.
110 A. Birch, *The Economic History of the British Iron and Steel Industry, 1784–1879*, 1967, p. 184; *Neilson v. Baird*, 1843, *Newtons Journal*, XXIII, 1843, Conjoined Series, p. 458.
111 Barlow, *op. cit.*, p. 21; Birch, *ibid.*, pp. 316–17; Birch claims that the merit of Heath's invention was exaggerated by contemporaries.
112 Morpeth Collectanea, M/16/B2; J. Raistrick to Magistrates of Hexham Sessions, M/16/B2, 30.7.1829; John Whinfield to T. Walker, 2/DE/44/10; J. Raistrick to Lord Daleval, 29.10.1803, 2/DE/33/10; Northumberland Record Office.

## Patentees, competition and the law

113 Barlow, *op. cit.*, p. 21.
114 *Select Committee on Patents*, Parl. Papers, III, 1829, p. 34. See also the evidence of A. Holdsworth, p. 131.
115 Felkin, *op. cit.*, pp. 205–48.
116 *Select Committee on Patents*, Parl. Papers, III, 1829, p. 144.
117 Felkin, *op. cit.*, p. 331.
118 Wedgwood MSS, J. Wedgwood to T. Bentley, 27.9.1768; and undated letter 1771 and 27.3.1772, John Rylands University Library of Manchester.
119 *Aris's Birmingham Gazette*, 20.6.1785, 27.6.1785. Both advertisements were on the front page. Users often signed agreements to inform the patentee of other infringers. See TR RAN CO5/5, Agreement between Ransomes and Busby, 1.12.1853, Reading University.
120 *Select Committee on Patents*, Parl. Papers, XVIII, 1851, p. 264.
121 Hindmarch, *op. cit.*, pp. 291–2.
122 Hartree MSS, Watt to McGregor, 30.10.1784, 119/M/30, Devon County Record Office; see also Boulton and Watt MSS, *Considerations upon the Measures to be adopted with Maberly*, Sept. 1796; Watt to Watt junior, 3.12.1798, Letter Book; Watt to Erskine, 16.12.1798, Letter Book, B.R.L.
123 Silberston and Taylor, *op. cit.*, p. 22. See also pp. 14–15, 21, 99–102.

# Conclusion

Two recent 'best guesses' at the sources of economic growth between 1700 and 1860 have emphasised (with varying force) the importance of technological change. Crafts estimates that the growth rate of total factor productivity increases throughout the eighteenth century. For the periods 1710–40 and 1740–80 the growth in the quantity of factor inputs was more important than productivity growth, but for the period 1780–1800 the accelerating growth of output owed slightly more to the growth of productivity than to capital, labour and land.[1] A similar estimate for the period 1780–1860 has been made by McCloskey, who argues that the steady increase in income per head (£11 in 1780, £28 in 1860) was attributable largely to changes on the supply side, especially in technology and the 'extraordinary flowering of ingenuity'.[2] In short, during the industrial revolution a given bundle of factor inputs produced progressively more output.

Both these studies have added substantial statistical weight to an old interpretation, and reassured a generation of schoolchildren! The purpose of the present study, however, has not been to assess the significances of technological change as an explanation of growth between 1750 and 1850, but to examine one neglected aspect of this problem: the role of inventive activity and the institutional framework within which it took place, namely the patent system.

The patent system never evoked the same sense of excitement as that associated with the development of technical innovation. In fact, many contemporaries thought the subject extremely dull but, equally, there is very little doubt that they considered patents to be a crucially important element in the process of technological change and economic growth. The existence of the patent system was, as far as they were concerned, one important contributory factor to the growing belief in economic progress. This is reflected in the arguments which were used to justify patents, and in the number of patents which were registered. Although there remains some doubt about the use of patents as an index of inventive activity during the industrial revolution (and after), a very large proportion of inventors did protect their ideas, and very few important inventions by-passed the system. For inventors, patents were effectively the only means by which they could appropriate a sufficient return for their effort. Many, of course, failed to reap any reward, but the right to exclude others from costlessly sharing in the benefits of their output was considered to be vital in a world which was becoming increasingly competitive. Patents provided security in an exceptionally risky activity, and protected inventors in their dealings with the users of inventive output.

The widespread use of patents to protect inventive output indicates that inventors were motivated by the expectation of making profits. Inventing was neither an altruistic nor a random activity, but an essentially economic one. The changing direction of inventive output between 1750 and 1850 confirms that inventors allocated their resources to take advantage of the differences between the market rate and the expected rate of return; the growing number of multiple inventors indicates that many anticipated the increasing importance of technology, and the gains which could be made through specialisation; the diversification of invention portfolios suggests also that inventors were conscious of the benefits of spreading risk; and, finally, the well developed trade in inventions shows that inventors were sensitive to the idea that users must be convinced of the benefits of altering the routine means of producing goods. The market for inventive output was not simple or deterministic. Inventors frequently had to market their output and persuade users of its value. Many were unsuccessful, but this should not disguise the fact that this is what they tried to do. England during the industrial revolution may well have been, as Brunel alleged, the best market for invention,[3] but this too should not lead to the conclusion that inventors simply responded to user demand. In large part, inventors had to create a demand for their output, and to do that they needed to protect their own property and offer users equivalent security. For this reason alone, patents were decisively important.

If patents were widely considered to be the most appropriate way to protect and reward inventors, and to ensure that economic progress was not constrained by the lack of 'flowering ingenuity', there is ample evidence that the patent system did not operate as effectively or as perfectly as many wished. The patent reform movement, which had a fairly long if discontinuous history, complained of a number of weaknesses in the administration of the system; the procedure of granting patents was absurdly cumbersome, and provided plenty of opportunity for cheating; the cost of patenting was considered as an unnecessary tax on ingenuity; and, most important of all, the poorly defined and uncertain nature of the law down to the 1830s tended to erode the value of patent property, and consequently tended to shift the terms of trade against the inventor. All these factors, in varying degrees, served to reduce the expected rate of return on inventive output. In practice, things were not always as bad as was imagined, although there is very little doubt that the inventor's perception of the problem was what really counted. If inventors firmly believed that patent protection was of little value then one would expect that they would either stop inventing, or seek other means, such as secrecy, to try and appropriate the rewards of their labour.

There is, however, very little evidence to suggest that this happened. The synchronisation of patent activity and the trade cycle in the first half of the nineteenth century suggests that underlying economic factors were more important. Secrecy, moreover, was not costless, and inventors, who were typically risk takers, rarely expected to end up in the courts. The adverse decisions of some judges obviously weakened the inventor's bargaining position *vis-à-vis* users, but despite the unpredictability of the courts, patents were still considered the only effective method of protecting inventive output, and they were worth having even when the costs involved were higher than expected. Inventors simply had to make the best of an imperfect industrial world.

Ironically, the imperfect nature of the patent system during the industrial revolution may in practice perhaps have approached the ideal. A system which efficiently protects inventors in every sense will slow down the rate of innovation and imitation, because the terms of trade tend to move in favour of the inventor. If the system is too imperfect in practice, the terms of trade will tend to move against him, and this might reduce the supply of inventive output and slow down the rate of technological change, *or* leave the supply of inventions unchanged, and quicken the speed of innovation and imitation. At any rate, the outcome is uncertain. On the other hand, a slightly imperfect system, where patents still provide some security for the inventor, or a degree of protection at least in excess of the next best alternative, will tend to accelerate the process of innovation and technological change. Of the three possibilities this would be the most conducive to economic growth.

Alternatively, if inventors optimistically overestimated the slowness of innovation and imitation (that is, they expected the diffusion lag to be longer than it actually was), then technological change would occur more rapidly compared with a situation in which inventors were pessimistic about the speed of diffusion. If patents create in inventors 'the socially wholesome illusion'[4] that the system is more perfect than it actually is, then the patent system may produce the best outcome: a relatively high rate of inventive activity and a relatively high rate of diffusion.

There is little evidence to support the 'socially wholesome illusion' effect, although it cannot be entirely discounted. Inventors, after all, were not always able to detect infringements and may have remained ignorant of the fact for some time, thus nourishing an optimistic estimate of the slowness of diffusion. In some cases, though, they were prepared to accelerate the adoption of their inventions to maximise royalties, and even if they were not, the growth of patent agents tended to keep them more and more informed of what was going on in the market. As is clear, there are some very complicated issues here.

If anything, it was the imperfect nature of the system which, paradoxically, created something close to the ideal. Inventors clearly realised that in practice it reduced the value of their output, but since they could do little about this for most of the period they were forced to live with the world as it was, and hope for better things to come. So long as patents provided a degree of protection over and above the next best alternative, as in fact appears to have been the case, it paid inventors to continue using them as the 'shield against the darts of [their] foes'. By accident, then, the imperfect nature of the patent system was in all probability the most appropriate for the economy as a whole for a long period of the industrial revolution.

## Notes

1 N. C. R. Crafts, 'The eighteenth century: a survey', in *The Economic History of Britain since 1700*, I, *1700–1860*, Cambridge, 1981, R. Floud and D. N. McCloskey, eds., pp. 1–16.
2 D. N. McCloskey, 'The industrial revolution, 1780–1860: a survey', in *ibid.*, pp. 103–27; see also G. N. von Tunzelman, 'Technical progress during the industrial revolution', pp. 143–63.
3 *Select Committee on Patents*, Parl. Papers, XVIII, 1851, p. 489.
4 F. Machlup, *The Economics of Sellers' Competition*, Baltimore, 1952. Machlup is concerned here with the innovator, but the argument can be applied to inventors, pp. 555–6.

# Appendix A
# The direction of patent activity, ranked by process, 1750–1851

| Rank | Process | Patents | % of total patents |
|---|---|---|---|
| 1. | Steam | 984 | 5·75 |
| 2. | Hydraulics | 750 | 4·38 |
| 3. | Spinning | 715 | 4·18 |
| 4. | Fuel | 714 | 4·17 |
| 5. | Carriages | 586 | 3·42 |
| 6. | Weaving | 502 | 2·93 |
| 7. | Bleaching | 479 | 2·80 |
| 8. | Non-ferrous metal | 474 | 2·77 |
| 9. | Marine propulsion | 461 | 2·69 |
| 10. | Acids, oxides | 454 | 2·65 |
| 11. | Raising and lowering | 392 | 2·29 |
| 12. | Oils and fats | 347 | 2·02 |
| 13. | Lace | 340 | 1·98 |
| 14. | Gas | 326 | 1·90 |
| 15. | Railway carriages | 325 | 1·90 |
| 16. | Finishing | 310 | 1·81 |
| 17. | Medical surgery | 288 | 1·68 |
| 18. | Ships | 287 | 1·67 |
| 19. | Iron and steel | 285 | 1·66 |
| 20. | Firearms and ammunition | 274 | 1·60 |
| 21. | Brewing | 273 | 1·59 |
| 22. | Gas engines | 267 | 1·56 |
| 23. | Musical instruments | 255 | 1·49 |
| 24. | Lamps | 240 | 1·40 |
| 25. | Letterpress printing | 234 | 1·36 |
| 26. | Agriculture I | 229 | 1·33 |
| 27. | Railways | 227 | 1·32 |
| 28. | Furniture | 219 | 1·28 |
| 29. | Optical instruments | 214 | 1·25 |
| 30. | Cooking | 199 | 1·16 |
| 31. | Electricity | 194 | 1·13 |
| 32. | Agriculture II | 177 | 1·03 |

| | | | |
|---|---|---|---|
| 33. | Paper | 171 | 0·99 |
| 34. | Fastening apparel | 171 | 0·99 |
| 35. | Sugar | 169 | 0·98 |
| 36. | Bricks and tiles | 153 | 0·89 |
| 37. | Paints | 150 | 0·87 |
| 38. | Saddlery | 146 | 0·85 |
| 39. | Body apparel | 142 | 0·83 |
| 40. | Artistic instruments | 131 | 0·76 |
| 41. | Cutting paper | 130 | 0·76 |
| 42. | Watches, clocks | 129 | 0·75 |
| 43. | Pottery | 128 | 0·74 |
| 44. | Nails | 127 | 0·74 |
| 45. | Harbours | 127 | 0·74 |
| 46. | Head apparel | 125 | 0·73 |
| 47. | Locks | 124 | 0·72 |
| 48. | Foot apparel | 120 | 0·70 |
| 49. | Grinding corn | 115 | 0·67 |
| 50. | Roads | 113 | 0·66 |
| 51. | Mining | 112 | 0·65 |
| 52. | Metal pipes | 110 | 0·64 |
| 53. | Masts, sails | 100 | 0·58 |
| 54. | Fire engines | 98 | 0·57 |
| 55. | Ventilation | 98 | 0·57 |
| 56. | Steering vessels | 95 | 0·55 |
| 57. | Artificial leather | 90 | 0·52 |
| 58. | Bottling liquids | 89 | 0·52 |
| 59. | Drains and sewers | 88 | 0·51 |
| 60. | Plating metals | 88 | 0·51 |
| 61. | Starch, gum | 87 | 0·50 |
| 62. | Aids to locomotives | 81 | 0·47 |
| 63. | Electricity and magnetism | 74 | 0·43 |
| 64. | Purifying water | 74 | 0·43 |
| 65. | India-rubber | 74 | 0·43 |
| 66. | Skins and hides | 67 | 0·39 |
| 67. | Brushing and sweeping | 67 | 0·39 |
| 68. | Artistic instruments | 67 | 0·39 |
| 69. | Unfermented beer | 66 | 0·38 |
| 70. | Steam culture | 65 | 0·38 |
| 71. | Food preservation | 61 | 0·35 |
| 72. | Farriery | 59 | 0·34 |
| 73. | Hinges | 55 | 0·32 |
| 74. | Bridges | 55 | 0·32 |
| 75. | Water closets | 54 | 0·31 |
| 76. | Washing machines | 54 | 0·31 |
| 77. | Manure | 48 | 0·28 |
| 78. | Agriculture III | 48 | 0·28 |
| 79. | Umbrellas | 44 | 0·25 |
| 80. | Needles and pins | 44 | 0·25 |
| 81. | Chains | 38 | 0·22 |

# Appendix A

| 82. | Tea and coffee | 34 | 0·19 |
|---|---|---|---|
| 83. | Tobacco | 34 | 0·19 |
| 84. | Books | 33 | 0·19 |
| 85. | Anchors | 29 | 0·16 |
| 86. | Sewing and embroidery | 29 | 0·16 |
| 87. | Trunks and baggage | 29 | 0·16 |
| 88. | Casks, barrels | 28 | 0·16 |
| 89. | Railway signals | 26 | 0·15 |
| 90. | Toys | 24 | 0·14 |
| 91. | Electricity and magnetism | 19 | 0·11 |
| 92. | Milking and churning | 17 | 0·09 |
| 93. | Electricity and magnetism | 15 | 0·08 |
| 94. | Photographs | 12 | 0·07 |
| 95. | Ice making | 12 | 0·07 |
| 96. | Aeronautics | 10 | 0·05 |
| 97. | Safes | 8 | 0·04 |
|  | *Total* | 17,101 |  |

*Source: Abridgements of Specifications*

During the industrial revolution, patent activity varied between industries and within industries over time, but no particular industry was especially significant in terms of taking out patent protection. Since 1945 electrical engineering, chemicals and pharmaceuticals have together consistently accounted for approximately 60% of all patents granted in the United Kingdom, but patenting was fairly evenly spread between 1750 and 1850. During the industrial revolution patent activity was not dominated by one or two industries.

# Appendix B
# The rate of patent survival for the United Kingdom, 1852–76

| Year | Patents sealed | % patents in force at the end of: 3 years | 7 years | 14 years |
|---|---|---|---|---|
| 1852 | 914 | 97·4 | 33·9 | 11·1 |
| 1853 | 2187 | 96·6 | 28·3 | 9·3 |
| 1854 | 1878 | 96·4 | 27·3 | 7·4 |
| 1855 | 2046 | 97·4 | 26·9 | 9·5 |
| 1856 | 2094 | 97·7 | 27·3 | 10·2 |
| 1857 | 2028 | 97·4 | 28·7 | 10·8 |
| 1858 | 1954 | 98·4 | 27·6 | 10·0 |
| 1859 | 1977 | 98·2 | 27·4 | 10·9 |
| 1860 | 2063 | 97·7 | 28·0 | 9·4 |
| 1861 | 2047 | 98·2 | 28·0 | 8·7 |
| 1862 | 2191 | 98·4 | 29·4 | 9·7 |
| 1863 | 2094 | 98·6 | 30·1 | 10·2 |
| 1864 | 2024 | 98·9 | 27·1 | 8·7 |
| 1865 | 2186 | 97·8 | 26·6 | 8·8 |
| 1866 | 2124 | 98·8 | 27·0 | 10·6 |
| 1867 | 2284 | 98·6 | 27·1 | 11·3 |
| 1868 | 2490 | 98·6 | 29·2 | 10·9 |
| 1869 | 2407 | 98·2 | 33·1 | 12·4 |
| 1870 | 2180 | 98·1 | 33·8 | 12·8 |
| 1871 | 2370 | 98·6 | 34·5 | 12·9 |
| 1872 | 2771 | 98·6 | 30·7 | 10·5 |
| 1873 | 2974 | 97·7 | 28·7 | 9·4 |
| 1874 | 3162 | 98·1 | 30·1 | 9·5 |
| 1875 | 3112 | 97·9 | 28·7 | 9·4 |
| 1876 | 3435 | 98·0 | 27·5 | 9·9 |

Source: *Eighteenth Report of Controller-General of Patents, Designs and Trade Marks* (Parl. Papers, XXIII, 1901).

# Appendix C
# The potential number of English patents in force, 1750–1851

| Date | Patents | Date | Patents | Date | Patents | Date | Patents | Date | Patents |
|---|---|---|---|---|---|---|---|---|---|
| 1750 | 102 | 1771 | 281 | 1792 | 724 | 1813 | 1396 | 1834 | 2230 |
| 1751 | 106 | 1772 | 296 | 1793 | 730 | 1814 | 1396 | 1835 | 2352 |
| 1752 | 107 | 1773 | 315 | 1794 | 752 | 1815 | 1394 | 1836 | 2535 |
| 1753 | 117 | 1774 | 336 | 1795 | 769 | 1816 | 1405 | 1837 | 2653 |
| 1754 | 122 | 1775 | 346 | 1796 | 805 | 1817 | 1435 | 1838 | 2867 |
| 1755 | 126 | 1776 | 359 | 1797 | 795 | 1818 | 1507 | 1839 | 3028 |
| 1756 | 123 | 1777 | 372 | 1798 | 826 | 1819 | 1513 | 1840 | 3327 |
| 1757 | 125 | 1778 | 384 | 1799 | 847 | 1820 | 1511 | 1841 | 3617 |
| 1758 | 122 | 1779 | 407 | 1800 | 883 | 1821 | 1526 | 1842 | 3834 |
| 1759 | 128 | 1780 | 409 | 1801 | 932 | 1822 | 1544 | 1843 | 4124 |
| 1760 | 138 | 1781 | 420 | 1802 | 997 | 1823 | 1581 | 1844 | 4394 |
| 1761 | 139 | 1782 | 436 | 1803 | 1027 | 1824 | 1653 | 1845 | 4815 |
| 1762 | 145 | 1783 | 464 | 1804 | 1019 | 1825 | 1788 | 1846 | 5161 |
| 1763 | 152 | 1784 | 480 | 1805 | 1057 | 1826 | 1811 | 1847 | 5474 |
| 1764 | 163 | 1785 | 519 | 1806 | 1071 | 1827 | 1830 | 1848 | 5655 |
| 1765 | 169 | 1786 | 550 | 1807 | 1122 | 1828 | 1888 | 1849 | 5938 |
| 1766 | 193 | 1787 | 576 | 1808 | 1162 | 1829 | 1916 | 1850 | 6155 |
| 1767 | 203 | 1788 | 583 | 1809 | 1212 | 1830 | 1978 | 1851 | 6354 |
| 1768 | 217 | 1789 | 606 | 1810 | 1245 | 1831 | 2026 | | |
| 1769 | 241 | 1790 | 645 | 1811 | 1306 | 1832 | 2041 | | |
| 1770 | 268 | 1791 | 669 | 1812 | 1347 | 1833 | 2120 | | |

*Source.* B. R. Mitchell and P. Deane, *Abstract of British Historical Statistics* (1962), pp. 268–9.

This table is constructed by adding the *net* change in patents every year, and it is important to note that the data show only the *potential* number of patents in force. Although inventors were still able to protect their patents by legal action at any time during the fourteen years of the patent's validity, many patents were probably discarded after two or three years. Before 1852 there are no data on the number of patents which were unused and it is, therefore, impossible to calculate exactly the *effective* patent population.

# Bibliography

**I. Manuscript sources**
*Berkshire County Record Office.* Hartley MSS.
*Birmingham Assay Office.* Boulton and Watt MSS.
*Birmingham Reference Library.* Boulton and Watt MSS.
*Birmingham University Library.* L. add. 701.
*British Museum.* Add. MSS. 27,807 Place Papers. Add. MSS. 38,345 Liverpool Papers. Egerton MSS. Kenyon MSS. Neilson *v.* Harford.
*Brotherton Library.* Holden MSS. Marriner MSS. Marshall MSS.
*Devon County Record Office.* Hartree MSS. Mason Tucker MSS.
*Derby Public Library.* Pares MSS. Strutt MSS.
*Gloucestershire County Record Office.* Mushet MSS.
*House of Lords Record Office.* Appendix to Public Petitions. Minutes of Evidence taken before the Lords Select Committee on Patents Amendment Bill.
*John Rylands University Library of Manchester.* Wedgwood MSS.
*Lancashire County Record Office.* Houghton, Craven MSS.
*Law Society.* Minute Books.
*Leeds City Library.* Bairstow MSS.
*Lincolnshire County Record Office.* Langton MSS.
*London University Library.* Raistrick MSS.
*Manchester Central Reference Library.* An Appreciation of R. B. Prosser. Diary of Samuel Taylor. Proceedings of the Manchester Chamber of Commerce. Stubs Wood MSS.
*National Library of Ireland.* Burke MSS.
*Northumberland County Record Office.* Morpeth Collectanea.
*Nottingham University Library.* Boden MSS.
*Public Record Office.* Board of Trade. Granville Papers. Home Office.
*Reading University.* Budding MSS.
*Redditch Public Library.* Gutch MSS.
*Royal Society of Arts.* Guard Books. R.S.A. Council Minutes. R.S.A. MSS.
*Salford University Library.* Badnell MSS.
*Scottish National Library.* Tweeddale MSS.
*Scottish Record Office.* Dundonald MSS.
*Staffordshire County Record Office.* Dartmouth MSS. Russell MSS.
*University College, London.* Brougham MSS. Chadwick MSS.
*University of Strathclyde.* James Young MSS.
*Wedgwood Museum.* Wedgwood MSS.

*Worcester County Record Office.* Gutch MSS.
*Private collection.* Spencer Moulton MSS.

## II. Parliamentary papers
*Hansard's Parliamentary Debates*, New and Third Series

PP. 1801 (55) III. Report from the Select Committee to whom the Petition of Matthias Koop, respecting his invention for making Paper from various refuse materials, was referred.

PP. 1809 (245) III. Report from the Select Committee to whom the Petition of Thomas Earnshaw, of High Holborn, in the county of Middlesex, was referred.

PP. 1810–11 (230) II. Report from the Select Committee to whom the Petition of Henry Greathead, of South Shields, in the county of Durham, was referred.

PP. 1812 (118) II. Report from the Select Committee to whom the Petition of Coningsby Cort, eldest son of the late Mr Henry Cort, of Gosport, in the County of Southampton, iron manufacturer, on behalf of the Petitioner and family of the said Henry Cort, was referred.

PP. 1812 (126) II. Report from the Select Committee to whom the Petition of Samuel Crompton, of Bolton-on-le-Moors, in the county of Lancaster, cotton-spinner, was referred.

PP. 1812–13 (67) III. Report from the Select Committee to whom the Petition of James Lee, of Frizo Water House, Enfield Wash, in the county of Middlesex, respecting a certain invention for preparing Hemp and Flax, was referred.

PP. 1817 (311) III. Report from the Select Committee to whom the Petition of Samuel Hill and William Bundy, and also the Petition of James Lee, relating to Machinery for Manufacturing of Flax and Hemp, was referred.

PP. 1818 (328) III. Report from the Select Committee to whom the Petition of John Leigh Bradbury, of Manchester, in the county of Lancaster, relative to Machinery for Engraving and Etching, was referred.

PP. 1824 (51) V. Report from the Select Committee appointed to inquire into the state of the law ... respecting Artisans ... the exportation of Tools and Machinery; and ... to the Combination of Workmen ... to regulate their Wages and Hours of Working.

PP. 1825 (504) V. Report from the Select Committee appointed to inquire into the state of the Law and its consequences, respecting the Exportation of Tools and Machinery.

PP. 1826 (216) XXI. Expenses incurred in taking out a Patent in England, Scotland and Ireland.

PP. 1826 (270) XXIII. Expenses incurred in taking out a Patent in Scotland.

PP. 1829 (332) III. Report from the Select Committee appointed to inquire into the present state of the Law and Practice relative to the granting of Patents for Invention.

PP. 1831–32 (380) V. Report from the Select Committee appointed to consider how far it is expedient to extend the patent granted for Morton's Slip.

PP. 1833 (690) VI. Report from the Select Committee appointed to inquire

## Bibliography

into the present state of Manufacturers, Commerce and Shipping in the United Kingdom.

PP. 1834 (483) XI. Report from the Select Committee on Mr Goldsworth Gurney's Case, as set forth in his Petition.

PP. 1835 (373) XIII. Report from the Select Committee appointed to inquire into the case of Mr Goldsworthy Gurney.

PP. 1835 (367) XLVIII. Report from the Select Committee appointed to investigate Mr Kyan's Patent for the prevention of the Dry Rot.

PP. 1837 (351) XX. Report from the Select Committee to whom the Petition of Henry and Sealy Fourdrinier was referred.

PP. 1837 (405) XX. Report on the re-committed Report from the Select Committee on Fourdrinier's Patent.

PP. 1837 (337) XXXIX. Minute of the Lords of the Committee of Privy Council for Trade, on granting Letters Patent.

PP. 1837–38 (278) XX. Reports from the Select Committee appointed to inquire into the present Rates and modes of Charging Postage ...

PP. 1837–38 (729) XXXVI. Fees payable on taking out a Patent of Invention in the United Kingdom; also for a Patent by more than One Person.

PP. 1839 (172) XXX. Fees and other Expenses payable on taking out a Patent of Invention for England, Scotland and Ireland.

PP. 1840 (155) XXXIX. Cases in which the Judicial Committee of the Privy Council have reported upon Petitions of persons having obtained Letters Patent ...

PP. 1841 (201) VII. Report from the Select Committee appointed to inquire into the operation of the existing Laws affecting the Exportation of Machinery.

PP. 1849 (576) XVIII. Report from the Select Committee on the School of Design.

PP. 1849 (1099) XXII. Report of the Committee (appointed by the Lords of the Treasury) on the Signet and Privy Seal Offices.

PP. 1849 (23) XLV. Number of Letters Patent for Inventions, sealed in each of the Ten Years, 1838–1848.

PP. 1851 (145) XVIII. Minutes of Evidence taken before the Select Committee of the House of Lords appointed to consider the Bill to extend the provisions of the Design Act, 1850, and to give protection from Piracy to persons exhibiting Inventions in the Exhibition of 1851.

PP. 1851 (486) XVIII. Report and Minutes of Evidence taken before the Select Committee of the House of Lords appointed to consider of the Bills for the amendment of the Law touching Letters Patent for Inventions with Appendix and Index.

PP. 1851 (509) XVIII. Report from the Select Committee on the Law of Partnership.

PP. 1864 (3419) XXIX. Report of the Commissioners appointed to inquire into the working of the law relating to Letters Patent for Invention.

PP. 1871 (368) X. Report from the Select Committee appointed to inquire into the Law and practice, and the effect of grants of letters patent for invention.

PP. 1872 (193) XI. Report from the Select Committee ... on the law and practice ... of letters patent for invention.

PP. 1894 (235) XIV. Special Report from the Select Committee on the Patent Agents Bill.
*House of Lords Journal*
*Journal of the House of Commons*

### III. Contemporary newspapers and journals
Aris's Birmingham Gazette
Derby Mercury
Manchester Guardian
Manchester Mercury
Scotsman
The Times

British Association for the Advancement of Science
Commissioners of Patents' Journal
Economist
Journal of the Society of Arts
Journal of the Statistical Society of London
Newtons Journal
Mechanics Magazine
Patent Journal
Quarterly Review
Rees Encyclopaedia
Repertory of Arts
Spectator
The Engineer
Transactions of the Institute of Patent Agents
Transactions of the Society of Arts
Westminster Review

### IV. Contemporary printed works (pre-1914)
Anderson J. *Observations on the Means of Exciting a Spirit of National Industry* (Edinburgh, 1777).
Babbage C. *On the Economy of Machinery and Manufactures* (London, 1832).
Babbage C. *Reflections on the Decline of Science in England and on some of its Causes* (London, 1830).
Bessemer H. *Autobiography* (London, 1905).
Billing S. and Prince A. *The Law and Practice of Patents and Registration of Designs* (London, 1845).
Bishop J. L. *A History of American Manufactures, 1608–1860* (Philadelphia, 1868, 3rd ed.).
Bramah J. *A Letter to the Right Hon. Sir James Eyre, L.C.J.C.C.P. on the subject of the cause Boulton and Watt v. Hornblower and Maberley for the infringement on Mr Watt's patent for an Improvement on the Steam Engine* (London, 1797).
Bramwell F. J. *The Expediency of Protection for Invention* (London, 1875).
Brunel I. *The Life of Isambard Kingdom Brunel* (London, 1870).
Burnley J. *The History of Wool and Woolcombing* (London, 1889).

## Bibliography 215

Campbell, J. L. *The Lives of the Chief Justices of England* (London, 1874, vol. IV).
Campin F. W. *Law of Patents for Invention* (London, 1869).
Carpmael W. *Law Reports of Patent Cases* (London, 1843–[52], 3 vols.).
Carpmael W. *The Law of Patents* (London, 1832).
Chitty J. *A Treatise on the Law of Commerce, Manufactures and Contracts* (London, 1820–24, vol. 1).
Clennell J. *Thoughts on the expediency of disclosing the process of manufactures* (London, 1807).
Cole H. *Fifty Years of the Public Work of Sir Henry Cole, K.C.B.* (London, 1884, 2 vols.).
Collier J. D. *An Essay on the Law of Patents* (London, 1803).
Colquhoun P. *Treatise on the Wealth, Power, and Resources of the British Empire* (London, 1815, 2nd ed.).
Coryton J. *A Treatise on the Law of Letters Patent* (London, 1855).
Davenport-Hill R. and F. *A Memoir of Matthew Davenport-Hill* (London, 1878).
Davies J. *Patent Cases* (London, 1816).
Dircks H. *Inventors and Invention* (London, 1867).
Dircks H. *Patent Monopoly as affecting the encouragement, improvement and progress of sciences, arts and manufactures* (London, 1869).
Donkin H. I. *Bryan Donkin and Co. Notes of History of an Engineering Firm during the Last Century, 1803–1903* (London, 1912).
Drewry C. S. *Observations on the Defects of the Law of Patents* (London, 1863).
Drewry C. S. *The Patent-Law Amendment Act* (London, 1837).
Edwards F. *On Letters Patent for Invention* (London, 1865).
Farrer T. H. *The State in Relation to Trade* (London, 1883).
Felkin W. *History of the Machine-Wrought Hosiery and Lace Manufactures* (Newton Abbot, 1967 ed.).
Fitzmaurice E. *The Life of Granville, George Leveson Gower, 2nd Earl Granville* (London, 1903).
Gill T. *Technological Repository* (London, 1822).
Godson R. *A Practical Treatise on the Law of Patents* (London, 1823).
Godson R. *A Supplement to a Practical Treatise on the Law of Patents for Invention* (London, 1832).
Godson R. *A Practical Treatise on the Law of Patents* (London, 1840, 2nd ed.).
Gordon J. W. *Monopolies by Patents* (London, 1897).
Hackwood F. W. *Wednesbury Workshop* (Wednesbury, 1889).
Hancock T. *Personal Narrative of the Origin and Process of the Caoutchouc or India Rubber Manufacture in England* (London, 1857).
Hands W. *The Law and Practice of Patents for Invention* (London, 1806).
Hindmarch W. *Law and Practice of Letters Patent for Invention* (London, 1848).
Hindmarch W. *Patent Laws of this country; suggestions for the Reform of them* (London, 1851).
Holroyd E. *A Practical Treatise on the Law of Patents for Invention* (London, 1830).

# Bibliography

James J. *History of the Worsted Manufacture in England* (London, 1857).
Kendrick W. *An Address to the Artists and Manufacturers of Great Britain Respecting an Application for the Encouragement of New Discoveries in the Useful Arts* (London, 1774).
Langford J. A. *Century of Birmingham Life, 1741–1841* (London, 1868).
Langton J. K. (ed.) *Memoirs of the Life and Correspondence of Henry Reeve* (London, 1898).
MacCleod H. D. *The Elements of Political Economy* (London, 1858).
Macintosh G. *Biographical Memoir: Charles Macintosh* (Glasgow, 1857).
McCulloch J. R. *Commercial Dictionary* (London, 1832).
McCulloch J. R. *The Literature of Political Economy* (London, 1845).
Macfie R. A. *Recent Discussions on the Abolition of Patents for Invention* (London, 1869).
Macfie R. A. *The Patent Question Under Free Trade* (London, 1875).
Macfie R. A. *The Patent Law Question* (London, 1863).
Meteyard E. *The Life of Josiah Wedgwood* (London, 1865 vol. 2).
Mill J. S. *Principles of Political Economy* (London, 1902 ed.).
Muirhead J. P. *The Life of J. Watt* (London, 1911).
Newton A. V. *Patent Law and Practice* (London, 1879).
Newton A. V. 'The patent agent's profession', *Transactions of the Institute of Patent Agents*, 1882–3.
Newton W. jnr. *In Memoriam: being a memoir of the late W. Newton* (London, 1861).
Norman J. P. *A Treatise on the Law and Practice relating to Letters Patent for Invention* (London, 1853).
Peacock J. *The Outline of a Plan for Establishing a United Company of British Manufacturers* (London, 1798).
Price W. H. *The English Patents of Monopoly* (London, 1906).
Prince A. *The Law and Practice of Patent and Registrations of Design* (London, 1845).
Prosser, R. B. *Birmingham Inventors and Inventions* (Birmingham, 1881).
Rae J. *New Principles of Political Economy* (Boston, 1834).
Ravenshear A. F. *The Industrial and Commercial Influence of the English Patent System* (London, 1908).
Roberts R. *Outlines of a Bill to Amend the Law for Granting Patents for Invention* (Manchester, 1833).
Roberts R. *Outlines of the Proposed Law of Patents for Mechanical Inventions* (Manchester, 1830).
Robertston J. C. *The Act to Consolidate and Amend the Laws Relating to Copyright and Design* (London, 1842).
Rogers J. E. T. 'On the rationale and working of the patent laws', *Journal of the Statistical Society of London*, 1863.
Smith A. *Lectures on Jurisprudence* (London, 1762–63; cited from Meek, R. L., Raphael, D. D., and Stein, P. G. eds, Oxford, 1978).
Smith A. *The Wealth of Nations* (London, 1776; cited from Campbell, R. H., Skinner, A. S., and Todd, W. B., eds, Oxford, 1976, 2 vols).
Smith J. W. *An Epitome on the Law Relating to Patents for Invention* (London, *1836)*.
Spence W. *Patentable Invention and Scientific Evidence* (London, 1851).

# Bibliography 217

Taussig F. W. *Inventors and Money-Makers* (New York, 1915).
Turner T. *Copyright in Design in Art and Manufactures* (London, 1848).
Turner T. *Counsel to Inventors* (London, 1850).
Turner T. *Remarks on the Rights of Property in Mechanical Invention with remarks to Registered Designs* (London, 1847).
Warden A. J. *The Linen Trade* (London, 1864).
Webster T. *Law and Practice of Letters Patent for Invention* (London, 1841).
Webster T. *On Property in Designs and Inventions in Arts and Manufactures* (London, 1853).
Webster T. *On the Amendment of the Law and Practice of Letters Patent for Inventions* (London, 1844, 2 vols).
Webster T. *Reports and Notes of Cases on Letters Patent for Inventions* (London, 1844, 2 vols).
Webster T. *The New Patent Law: its history, objects and provisions ... etc.* (London, 1852).
Wheeler J. *Manchester: its Political, Social and Commercial History* (Manchester, 1836).
Wise L. 'On patent agents: their profession considered as a necessity, with suggestions for reform', *Transactions of the Institute of Patent Agents*, 1885–86.
Wood T. H. *History of the Royal Society of Arts* (London, 1913).
Woodcroft B. *Alphabetical Index of Patentees of Invention* (London, 1854).
Woodcroft B. *Brief Biographies of Inventors of Machines for the Manufacture of Textile Fabrics* (London, 1863).
Woodcroft B. *Chronological Index of Patents of Invention* (London, 1854).
Woodcroft B. *Patents for Invention: Reference Index, 1617–1853* (London, 1855).
Wyatt W. H. *A Compendium of the Law of Patents for Invention* (London, 1826).
Wyndam Hulme E. 'On the consideration of the patent grant, past and present', *Law Quarterly Review*, XIII, 1897.
Wyndam Hulme E. 'On the history of patent law in the seventeenth and eighteenth centuries', *Law Quarterly Review*, XVIII, 1902.
Wyndam Hulme E. 'Privy Council law and practice of letters patent for invention from the Restoration to 1794', *Law Quarterly Review*, XXIII, 1907.
*A Memoir of the Life, Writings and Mechanical Inventions of Edmund Cartwright* (1971, ed.).
*Lectures delivered by Prof. Amos at London University on the Law relating to Patents* (London, 1835).
*Papers Relating to Mr Champion's Application to Parliament for an Extension of the Term of a Patent* (London, 1775).
*Speech by W A Mackinnon, MP on Letters Patent for Invention* (London, 1838).
*The Trial of a Cause instituted by R P Arden Esq. His Majesty's Attorney General, by a Writ of Scre Facias, to repeal a Patent Granted on the 16th December 1775 to Mr Richard Arkwright* (London, 1785).
*The Life of Lloyd, the first Lord Kenyon L.C.J. of England* (London, 1873).
*A Sketch of the Life and Character of Lord Kenyon* (London, 1802).

## V. Secondary printed works
*Books*

Aldcroft D. H. and Fearon P. (eds). *British Economic Fluctuations, 1790–1939* (London, 1972).

Allan D. G. C. *William Shipley, Founder of the Royal Society of Arts* (London, 1968).

Arrow K. 'Economic welfare and the allocation of resources for invention', in Rosenberg, N. (ed.), *The Economics of Technological Change* (London, 1971).

Ashton T. S. *Economic Fluctuations in England, 1700–1800* (Oxford, 1959).

Ashton T. S. *Iron and Steel in the Industrial Revolution* (Manchester, 1963, 3rd ed.).

Barker T. C. *The Glassmakers: Pilkington, 1826–1976* (London, 1977).

Barker T. C. and Harris J. R. *A Merseyside Town in the Industrial Revolution: St. Helens, 1750–1900* (London, 1959).

Beresford M. W. *The Leeds Chamber of Commerce* (Leeds, 1951).

Birch A. *The Economic History of the British Iron and Steel Industry, 1784–1879* (London, 1967).

Boehm K. and Silberston A. *The British Patent System: Administration* (Cambridge, 1967).

Bowden W. *Industrial Society in England towards the End of the Eighteenth Century* (New York, 1925).

Cameron H. C. *Samuel Crompton, 1753–1827* (London, 1951).

Carter C. F. and Williams B. R. *Investment in Innovation* (London, 1958).

Chapman S. D. *The Early Factory Masters* (Newton Abbot, 1967).

Church R. A. and Chapman S. D. 'Gravenor Henson and the making of the English working class', in Jones, E. L., and Mingay, G. E. (eds.), *Land, Labour and Population in the Industrial Revolution* (London, 1967).

Clow A. and N. L. *The Chemical Revolution: a contribution to social technology* (London, 1952).

Coleman D. C. *The British Paper Industry, 1495–1860* (Oxford, 1958).

Coleman, D. C. 'Textile growth', in Harte, N. B., and Ponting, K. G. (eds.), *Textile History and Economic History: Essays in Honour of Miss Julia de Lacy Mann* (Manchester, 1973).

Cottrell P. L. *Industrial finance, 1830–1914: the finance and organisation of English Manufacturing Industry* (London, 1980).

Crafts N. F. R. 'The eighteenth century: a survey', in Floud, R., and McCloskey, D. N. (eds.), *The Economic History of Britain since 1700*, vol. 1, *1700–1800* (Cambridge, 1981).

Cropp J. A. D., Harris D. S. and Stern E. S. *Trade in Innovation* (London, 1970).

Crouzet F. *Capital Formation in the Industrial Revolution* (London, 1972).

Crump W. B. (ed.). *Leeds Woollen Industry, 1780–1820* (Leeds, 1931).

Dickinson H. W. *Matthew Boulton* (Cambridge, 1937).

Dreyer J. S. (ed.). *Breadth and Depth in Economics* (San Diego, 1979).

Ede J. F. *History of Wednesbury* (London, 1962).

Elsas M. (ed.). *Iron in the Making: Dowlais Iron Co. Letters, 1782–1860* (Cardiff, 1960).

# Bibliography

Feinstein C. H. 'Capital formation in Great Britain', in Mathias, P., and Postan, M. M. (eds.). *Cambridge Economic History of Europe* (Cambridge, 1978, vol. VII).

Finer A. and Savage G. *J. Wedgwood: Selected Letters* (London, 1965).

Fitton R. S. and Wadsworth A. P. *The Strutts and the Arkwrights* (Manchester, 1964).

Floud R. and McCloskey D. N. (eds.). *The Economic History of Britain since 1700:* vol. 1: *1700–1860* (Cambridge, 1981).

Forwell W. R. (ed.). *Chartered Institute of Patent Agents: Informals: Collected Papers* (London, 1969).

Fox H. G. *Monopolies and Patents* (Toronto, 1947).

Freeman C. *The Economics of Industrial Innovation* (London, 1974).

Gauldie E. *The Dundee Textile Industry, 1790–1885* (Edinburgh, 1969).

Gayer A. D., Rostow W. W. and Schwartz A. J. *The Growth and Fluctuations of the British Economy, 1790–1850* (Oxford, 1953, 2 vols.).

Gilfillan S. C. *Sociology of Invention* (Chicago, 1935).

Gomme A. A. *Patents of Invention: Origins and Growth of the Patent System in Britain* (London, 1946).

Griliches Z. and Hurwicz L. (eds.). *Patents, Invention and Economic Change* (Harvard, 1972).

Harding A. *A Social History of English Law* (London, 1966).

Harding H. *Patent Office Centenary* (London, 1953).

Harte N. B. and Ponting K. G. (eds.). *Textile History and Economic History: Essays in Honour of Miss Julia de Lacy Mann* (Manchester, 1973).

Hartwell R. M. 'The Service Revolution', in Cipolla, C. M. (ed.). *The Fontana Economic History of Europe* (London, 1973, vol. 3).

Hatfield S. *Inventions and their Use in Science Today* (London, 1939).

Henderson W. O. (ed.). *J. C. Fischer and his Diary of Industrial England, 1814–1851* (London, 1966).

Henson G. *History of the Framework Knitters* (Newton Abbot, 1970 ed.).

Holderness B. A. 'Capital Formation in Agriculture', in Higgins, J. P. P., and Pollard, S. (eds.). *Aspects of Capital Investment in Great Britain, 1750–1850* (London, 1971).

Holdsworth W. *History of English Law* (London, 1922–66, vol. XI).

Hudson D. and Luckhurst K. W. *The Royal Society of Arts, 1754–1954* (London, 1954).

Jenkins D. T. *The West Riding Wool Textile Industry 1770–1835: A Study of Fixed Capital Formation* (Edington, Wiltshire, 1975).

Jeremy, D. J. *Transatlantic Industrial Revolution: The Diffusion of Textile Technologies between Britain and America, 1790–1830* (Cambridge, Massachusetts, 1981).

Johnson P. S. *The Economics of Invention and Innovation: with a case study of the development of the Hovercraft* (London, 1975).

Kaempffret W. B. *Invention and Society* (Chicago, 1930).

Kingston W. *Invention and Monopoly* (London, 1968).

Kuznets S. 'Inventive activity: problems of definition and measurement', in Nelson, R. R. (ed.), *The Rate and Direction of Inventive Activity* (Princeton, 1962).

Lamberton D. M. (ed.). *Economics of Information and Knowledge* (London, 1971).
Landes D. S. *The Unbound Prometheus* (Cambridge, 1970).
Lilly S. *Machines and History* (London, 1948 ed.).
McCloskey D. N. 'The industrial revolution, 1780–1860: a survey', in Floud, R., and McCloskey, D. N. (eds.), *The Economic History of Britain since 1700: Vol. 1: 1700–1860* (Cambridge, 1981).
Machlup F. *The Economics of Sellers' Competition* (Baltimore, 1952).
Machlup F. *The Production and Distribution of Knowledge in the United States* (Princeton, 1962).
McKie D. and Robinson E. *Partners in Science: Letters of James Watt and Joseph Black* (London, 1969).
de Lacy Mann J. *The Cloth Industry in the West of England from 1640 to 1880* (Oxford, 1971).
Mansfield E. A. *The Economics of Technological Change* (London, 1969).
Mansfield E. A. *Industrial Research and Technological Innovation* (London, 1968).
Mitchell B. R. and Deane P. *Abstract of British Historical Statistics* (Cambridge, 1962).
Musson A. E. and Robinson E. *James Watt and the Steam Revolution* (London, 1969).
Musson, A. E. *Science, Technology and Economic Growth in the Eighteenth Century* (London, 1972).
Musson A. E. and Robinson E. *Science and Technology in the Industrial Revolution* (Manchester, 1969).
Nelson R. R. (ed.). *The Rate and Direction of Inventive Activity* (Princeton, 1962).
Nelson R. R., Peck M. J. and Kalachek E. D. *Technology, Economic Growth and Public Policy* (Princeton, 1967).
Nordhaus W. D. *Invention, Growth and Welfare: a theoretical treatment of technological change* (Massachusetts, 1969).
Norris K. and Vaizey J. *The Economics of Research and Technology* (London, 1973).
North D. C. and Thomas R. P. *The Rise of the Western World: A New Economic History* (Cambridge, 1976).
Parker J. E. S. *The Economics of Innovation: the national and multinational enterprise in technological change* (London, 1974).
Payne P. L. *Rubber and Railways in the Nineteenth Century* (Liverpool, 1961).
Penrose E. *The Economics of the International Patent System* (Baltimore, 1951).
Pole W. (ed.). *The Life of Sir William Fairbairn ... etc.* (London, 1970 ed.).
Pollard S. *The Genesis of Modern Management* (London, 1965).
Prosi G. 'Socially optimal patent protection — another exercise in utopian economics?', in Dreyer, J. S. (ed.), *Breadth and Depth in Economics* (San Diego, 1979).
Redford A. *Manchester Merchants and Foreign Trade, 1794–1858* (Manchester, 1934).
Rimmer W. G. *Marshall's of Leeds: Flax-Spinners, 1788–1886* (Cambridge, 1960).

# Bibliography

Roll E. *An Early Experiment in Industrial Organization* (London, 1930).
Rosenberg, N. *Perspectives in Technology* (Cambridge, 1976).
Rosenberg N. *Technology and American Economic Growth* (New York, 1972).
Rossman J. *The Psychology of the Inventor* (Washington, D.C. 1931).
Salter W. E. G. *Productivity and Technical Change* (Cambridge, 1969).
Schiff E. *Industrialization without National Patents: The Netherlands, 1869–1912, Switzerland, 1850–1907* (Princeton, 1971).
Schmookler J. *Invention and Economic Growth* (Cambridge, Mass., 1966).
Schofield R. E. *The Lunar Society of Birmingham* (Oxford, 1964).
Schumpeter J. *Business Cycles* (New York, 1939, vol. 1).
Sigsworth E. M. 'Sir Isaac Holden, Bt: the first comber in Europe', in Harte, N. B., and Ponting, K. G. (eds.), *Textile History and Economic History: Essays in Honour of Miss Julia de Lacy Mann* (Manchester, 1973).
Smiles S. *Industrial Biography* (Newton Abbot, 1967 ed.).
Smiles S. *Lives of the Engineers: Early Engineering* (London, 1904).
Stamp Sir J. *Invention as an Economic Factor* (London, 1928).
Stamp Sir J. *Some Economic Factors in Modern Life* (London, 1929).
Stark W. (ed.). *J. Bentham's Economic Writings* (London, 1952).
Strassman W. P. *Risk and Technological Innovation* (Ithaca, 1959).
Taylor C. T. and Silberston A. *The Economic Impact of the Patent System: A Study of the British Experience* (Cambridge, 1973).
von Tunzelmann G. N. 'Technical progress during the industrial revolution', in Floud, R., and McCloskey, D. N. (eds.), *The Economic History of Britain since 1700:* vol. 1: *1700–1800* (Cambridge, 1981).
Usher A. P. *A History of Mechanical Inventions* (Cambridge, Mass. 1954).
Vaughan F. *Economics of the Patent System* (New York, 1925).
Woodruff W. *The Rise of the British Rubber Industry in the Nineteenth Century* (Liverpool, 1958).

*Articles*
Ashton T. S. 'The records of a pin manufactory', *Economica*, V, 1925.
Ashton T. S. 'Some statistics of the industrial revolution in Britain', *Manchester School*, XVI, 1948.
Bargar D. 'Matthew Boulton and the Birmingham petition of 1775', *William and Mary Quarterly*, XIII, 1956.
Barker T. C., Dickinson R. and Hardie D. W. F. 'The origins of the synthetic alkali industry in Britain', *Economica*, XXIII, 1956.
Bowden W. 'The influence of the manufacturer on some of the early policies of William Pitt', *American Historical Review*, XXIX, 1924.
Chapman S. D. 'Financial restraints on the growth of firms in the cotton industry', *Economic History Review*, XXXII, 1979.
Chapman S. D. 'The transition of the factory system in the Midlands cotton spinning industry', *Economic History Review*, XVIII, 1965.
Clow A. and N. L. 'Vitriol in the industrial revolution', *Economic History Review*, XV, 1945.
Crafts N. F. R. 'Industrial revolution in England and France: some thoughts on the question, why was England first', *Economic History Review*, XXI, 1972.

Cule J. E. 'Finance and industry in the eighteenth century: firm of Boulton and Watt', *Economic History*, IV, 1940.
Daff T. 'Patents as history', *Local Historian*, IX, 1970–71.
Dickinson H. W. 'Richard Roberts, his life and inventions', *Transactions of the Newcomen Society*, XXV, 1945–7.
Encel S. and Inglis A. 'Patents, invention, and economic progress', *Economic Record*, XLII, 1966.
Epstein R. C. 'Industrial inventions: heroic or systematic?', *Quarterly Journal of Economics*, XL, 1936.
Frumkin M. 'The origin of patents', *Journal of the Patent Office Society*, XXVII, 1945.
Gattrell V. A. C. 'Labour, power and the size of firms in Lancashire cotton in the second quarter of the nineteenth century', *Economic History Review*, XXX, 1977.
Getz L. 'A history of the patentee's obligation in Great Britain', *Journal of the Patent Office Society*, XLVI, 1964.
Gilfillan S. C. 'Invention as a factor in economic history', *Journal of Economic History*, V, 1945.
Gilfillan S. C. 'The prediction of technological change', *Review of Economic Statistics*, XXXIV, 1952.
Gomme A. A. 'Centenary of the Patent Office', *Transactions of the Newcomen Society*, XXVIII, 1951–3.
Hamilton W. 'Origin and early history of patents', *Journal of the Patent Office Society*, XVIII, 1936.
Johnson H. G. 'Patents and licences as stimuli to innovation', *Weltwirtschaftliches Archiv*, CXII, 1976.
Jones S. R. H. 'Price associations and competition in the British pin industry, 1814–1840', *Economic History Review*, XXVI, 1973.
Kennedy C. and Thirwall A. P. 'Surveys in applied economics: technology', *Economic Journal*, LXXXII, 1972.
Machlup F. *An economic review of the patent system*, Study No. 15 of the sub-committee on patents, trademarks and copyrights of the Committee on the Judiciary (U.S. Senate, 85th Congress, 2nd session, Washington, 1958).
Machlup F. and Penrose E. 'The patent controversy in the nineteenth century', *Journal of Economic History*, X, 1950.
McKendrick N. 'Wedgwood and Thomas Bentley: an inventor-entrepreneur partnership in the industrial revolution', *Transactions of the Royal Historical Society*, XIV, 1964.
McKendrick N. 'Josiah Wedgwood: an eighteenth-century entrepreneur in salesmanship and marketing techniques', *Economic History Review*, XXIII, 1960.
Maclaurin W. R. 'The sequence from invention to innovation and its relation to economic growth', *Quarterly Journal of Economics*, XVII, 1953.
Merton, R. K. 'Fluctuations in the rate of industrial inventions', *Quarterly Journal of Economics*, XLIX, 1934–5.
Nelson R. R. 'A survey of the economics of invention', *Journal of Business*, XXXII, 1959.
Nelson R. R. and Winter S. G. 'Neoclassical vs. evolutionary theories of

economic growth: Critique and prospectus', *Economic Journal*, LXXXIV, 1974.
Nordhaus W. D. 'The optimal life of a patent: reply', *American Economic Review*, LXII, 1972.
Norris J. M. 'Samuel Garbett and the early development of industrial lobbying in Great Britain', *Economic History Review*, X, 1957–8.
Plant A. 'The economic theory concerning patents for invention', *Economica*, I, 1934.
Pollard S. 'Capital accounting in the industrial revolution', *Yorkshire Bulletin of Economic and Social Research*, XV, 1963.
Pollard S. 'Fixed capital in the industrial revolution in Britain', *Journal of Economic History*, XXIV, 1964.
Post R. C. '"Liberalizers" versus "scientific men" in the antebellum Patent Office', *Technology and Culture*, XVII, 1976.
Prager J. 'Early growth and influence of intellectual property', *Journal of the Patent Office Society*, XLVI, 1964.
Prosi G. 'Patents and externalities', *Zeitschrift für Nationalökonomie*, XXXI, 1971.
Robinson E. 'Eighteenth century commerce and fashion: Matthew Boulton's marketing techniques', *Economic History Review*, XVI, 1963.
Robinson E. 'James Watt and the law of patents', *Technology and Culture*, XII, 1971.
Robinson E. 'Matthew Boulton and the art of parliamentary lobbying', *Historical Journal*, VII, 1964.
Rosenberg N. 'Science, invention and economic growth, *Economic Journal*, LXXXIV, 1974.
Rosenberg N. 'On technological expectations', *Economic Journal*, LXXXVI, 1976.
Ruttan V. W. 'Usher and Schumpeter on invention and innovation and technological change', *Quarterly Journal of Economics*, LXXIII, 1959.
Seaborne Davies D. 'Early history of English patent specifications', *Law Quarterly Review*, L, 1934.
Scherer F. M. 'Patent statistics as a measure of technical change', *Journal of Political Economy*, LXXVII, 1969.
Scherer H. 'The Mackintosh: the paternity of an invention', *Transactions of the Newcomen Society*, XXVIII, 1951–3.
Scherer F. M. 'Nordhaus's theory of optimal patent life: a geometric reinterpretation', *American Economic Review*, LXII, 1972.
Schmookler J. 'Economic sources of inventive activity', *Journal of Economic History*, XXII, 1962.
Schmookler J. 'Invention, innovation and competition', *Southern Economic Journal*, XX, 1954.
Schmookler J. 'Inventors past and present', *Review of Economics and Statistics*, XXXIX, 1957.
Schmookler J. 'The level of inventive activity', *Review of Economics and Statistics*, XXXVI, 1954.
Schmookler J. 'The utility of patent statistics', *Journal of the Patent Office Society*, XXXV, 1953.
Schmookler J. 'Interpretation of patent statistics', *Journal of the Patent Office Society*, XXXII, 1950.

Silberston A. 'The patent system', *Lloyds Bank Review*, LXXXIV, 1967.
Smith M. 'Patents for invention: the national and local picture', *Business History*, IV, 1961–2.
Solo C. S. 'Innovation in the capitalist process: a critique of the Schumpeterian theory', *Quarterly Journal of Economics*, LXV, 1951.
Tann J. 'Marketing methods in the international steam engine market: the case of Boulton and Watt', *Journal of Economic History*, XXXVIII, 1978.
Tann J. 'Richard Arkwright and technology', *History*, LVIII, 1973.
Tann J. 'The international diffusion of the Watt engine, 1775–1825', *Economic History Review*, XXXI, 1978.

## VI. Unpublished theses
Dutton H. I. *The Patent System and Inventive Activity during the Industrial Revolution, 1750–1852.* (Ph.D., London, 1981).
MacLeod C. *Patent for Invention and Technical Change in England, 1660–1753.* (Ph.D., Cambridge, 1982).
van Zyle Smit D. *The Social Creation of a Legal Reality: a Study of the Emergence and Acceptance of the British Patent System as a Legal Instrument for the Control of New Technology.* (Ph.D., Edinburgh, 1980).

# Index

Abbot, Francis, 45, 77
Abbot, Justice, 72–3, 74
Abbot, Mr., 87
Abramovitz, M., 13
agriculture, 11, 165, 176, 193
Albert, Charles, 95
Aldcroft, D. H., 196, 218
Alderson, Baron, 78
alkali manufacture, 139–9, 147, 148, 183
Allan, D. G. C., 32, 218
Allhusen, Christian, 94
Anderson, J., 214
Anstruther, J., 41
anti-patent movement, 17, 23–9, 59
Archer, Henry, 57–8
Argand, Aimé, 38, 40, 52
*Aris's Birmingham Gazette*, 201, 214
Arkwright, Richard, 26, 27, 28, 37, 39, 40, 69, 76, 111, 134, 137, 140, 153, 157, 160, 188, 189, 194, 217
Armstrong, William, 25
Arrow, K., 11
Art Protection Society, 58
*Artisan Journal*, 58
Ashton, T. S., 3, 10, 11, 52, 118, 143, 169, 196, 218, 221
Ashurst, Justice, 80
assignments, 93
Association of Patentees and Proprietors of Patent, 58
Association of Patentees for the Protection and Regulation of Patent Property, 21
Atkinson, John, 190
attitude of Judges, 76–81
Auckland, Lord, 152

Babbage, Charles, 26, 32, 70, 109, 119, 144, 152, 214
Baddeley, W., 87
Bainbridge, Mr., 77
Baird, William, 193
Bairstow, T. and M., 162
Banks, Joseph, 12
Barber, Alfred, 140
Bargar, D., 53, 169, 221
Barker, T. C., 143, 146, 147, 148, 172, 218, 221
Barlow, William, 79, 87, 145, 149, 170, 178, 193, 200, 201
Bates, J., 87
Batzel, V. M., 33
Beard, Mr., 140
Beaumont and Wackerbank, 160
Belfast, 59

Bentham, Jeremy, 19, 65
Benthamism, 65
Benson, Dr., 188
Bentley, Thomas, 184
Benyon, Bros., 161
Bernard, Mr., 41
Bernie, J. B., 165–8
Berry, Miles, 87, 96
Bessemer, Henry, 7, 12, 109, 119, 123, 125, 131, 143, 145, 159, 169, 172, 214
Bethall, J., 87
Bewley, Mr., 161
Billing, S., 214
Birch, A., 200, 218
Birkbeck, Dr., 43
Birkinshaw, John, 43
Birley, H. H., 161
Birmingham, 44, 46, 49, 59, 87, 88, 125, 126, 129, 194
Birmingham Chamber of Commerce, 46
Birmingham Patent Law Reform Association, 58
Bishop, J. L., 145, 214
Blackburn, 59, 62, 63
Bladen, W. W., 120
Blaug, M., 85
Bloxam and Co., 161
Blunt, R., 88
Board of Trade, 57, 58, 152, 163
Boards of Examiners, 45, 46, 48, 61, 92
Boards of Invention, 26
Bodmer, J. C., 184
Boehm, K., 1, 2, 10, 26, 32, 51, 110, 218
Bolton, 140
Boomer, James, 190
Booth, James, 42
Boulton, Matthew, 38, 40, 76, 136, 140, 151, 182, 183, 184
Bourboulon de Bonevil, A., 28
Bovill, G. H., 193
Bowden, W., 26, 32, 52, 118, 218
Bower and Bark, 129
Bowman, Robert, 134, 157
Bowring, Mr., 50
Boyman Boyman, Mr., 158
Boyson, R., 171
J. Boz and Co., 190
Bradbury, John Leigh, 128, 212
Bradford, 59, 63, 140
Bramah, Joseph, 7, 73, 82, 157, 214
Bramwell, F. J., 123, 142, 214
Brenner, J. F., 85
brick making, 164–8, 185
Bridges, Robert, 168

# Index

Bridport, 59
Brimscombe, 192
Bristol, 126
*British Association for the Advancement of Science*, 32, 68, 214
British Plate Glass Manufactory, 138
bronze powder, 125-6, 159
Brooman, R. A., 87
Brougham, Lord, 21, 47, 48-51, 57, 58, 60, 61, 62, 65, 84, 95, 155, 169
Brown, Lucy, 54
Brunel, I., 33, 214
Brunel, I. K., 25, 29, 33, 60, 166, 203
Brunel, Marc, 44, 45, 124, 159
Brush, Jarvis, 130
Bubble Act, 152
Buccleuch, Duke of, 137
Buchanan, J. M., 30
Buck, Mr., 189
Budding Edwin, 160
Buller, Justice, 22, 74, 75, 77, 82
Bundy, William, 127, 212
Burke, Edmund, 23, 31
Burn, Robert Scott, 57
Burnley, 140
Burnley, J., 146, 172, 214

calico printing, 90, 158
Campbell, J. L., 78, 82, 84, 215
Campin, F. W., 58, 60, 87, 215
candle industry, 131-2
carpet industry, 162
Carpmael, William, 22, 60, 61, 62, 63, 81, 82, 83, 84, 85, 87, 88, 90, 93, 97, 98, 99, 123, 143, 144, 151, 152, 155, 170, 172, 183, 184, 196, 197, 198, 215
Carter, C. F., 118, 218
Cartwright, Edmund, 27, 32, 41, 157, 180, 217
Case, Mr., 133
Chaloner, W. H., 173
Chamberlain, John, 24
Chance, J., 131, 134, 146
Chapman, S. D., 144, 146, 147, 156, 171, 199, 218, 221
chemical industry, 44, 110, 179
Chitty, J., 21, 28, 31, 215
Church, R. A., 53, 218
Clanricarde, Marquess, 43
Clapham, J. H., 103, 117
Clegg, Samuel, 45, 70, 154, 181
Clennell, J., 215
Clow, A. and N. L., 139, 147, 148, 218, 221
Cochrane, Lord, 90
Cochrane, W. A., 151
Cocker, Mr., 185
Cole, Henry, 60, 194, 215
Coleman, D. C., 121, 147, 171, 172, 173, 218
Collier, J. D., 31, 73, 82, 215
Collins, William, 125
collusion, 50, 187-94
Colquhoun, P., 32, 215
commercial committees, 36
Commissioners of Patents, 61
*Commissioners of Patents' Journal*, 143, 214
Committee for the Encouragement of Trade, 37
Committee of the Society of Arts for Legislative Recognition of the Rights of Inventors, 58, 60, 62
companies working patents, 164
competition, 179, 182-7, 195

Cook, Edward, 130
Cooper, H., 183, 184
copper smelting, 132
Cort, Henry, 153, 157, 212
Coryton, J. A., 84, 85, 96, 143, 215
cost of legal action, 179, 181-2, 195
Cottan and Hallam, 168
cotton industry, 36-7, 44, 114, 115, 122, 123, 134-5, 141, 156, 157-8, 188-9
Cottrell, P. L., 156, 170, 171, 173, 218
Counters, John, 141
court cases, 70-2
  *Arkwright* v. *Nightingale*, 80
  *Barber* v. *Grace*, 83
  *Beard* v. *Barber*, 148
  *Beaumont* v. *George*, 84, 172
  *Bloxam* v. *Else*, 173
  *Bovill* v. *Moore*, 83, 193
  *Chante* v. *Leese*, 146
  *Cornish* v. *Keene*, 84, 196
  *Crane* v. *Price*, 75, 85, 88
  *Crossley* v. *Beverly*, 81
  *Derosne* v. *Fairre*, 83
  *Fisher* v. *Oliver and Atkins*, 200
  *Hancock* v. *Bunsen*, 172
  *Hartley* v. *Hadland*, 146
  *Hayne* v. *Maltby*, 146
  *Hill* v. *Thompson*, 74, 83, 173, 196
  *Hornblower* v. *Boulton*, 20, 72, 77, 80
  *Horsehill and Iron Co.* v. *Neilson and Others*, 82
  *Huddart* v. *Grimshaw*, 84
  *Hullet* v. *Hague*, 84
  *Kay* v. *Marshall*, 84, 200
  *King* v. *Arkwright*, 22, 75
  *King* v. *Wheeler*, 82, 83
  *Lewis* v. *Davies*, 200
  *Lewis* v. *Marling*, 200
  *Liardet* v. *Johnson*, 22, 39, 75, 81
  *Marling* v. *Kirby*, 144
  *Minter* v. *Wells*, 173
  *Morgan* v. *Seaward*, 83, 85
  *Neilson* v. *Hartford*, 78, 84, 146, 147
  *Regina* v. *Cutler*, 85
  *Russell* v. *Cowley*, 84, 196
  *Turner* v. *Winter*, 84
  *Williams* v. *Williams*, 22
Coventry, 44
Cowcher, Kirby and Co., 128, 129
Cowper, C., 87
Crafts, N. F. R., 118, 202, 205, 218, 221
Crompton, Samuel, 12, 153, 212
Cropp, J. A. D., 145, 218
Crouch, R. L., 30
Crouzet, F., 153, 158, 169, 170, 171, 218
Crowder, Joseph, 42
Crump, W. B., 197, 218
Cubitt, William, 24, 29
Cule, J. E., 171, 172, 222
Cunningham, W., 103
custom and routine, 125, 130, 142, 150, 154, 203

Daff, T., 26, 32
Dance, Sir Charles, 162
Daniel, J. C., 181
Daniels, G. W., 188, 199
data problems, 6-8, 108, 150-1, 155-6
Davenport-Hill, Matthew, 29, 33, 215
Davies, J., 31, 83, 215

# Index

Davies, William, 192
Davis, R., 1, 11
Dean and Henderson, 165
Deane, P., 118, 195
De Girard, 189
Denison, E. B., 29
Denman, Lord, 67
Denman, Sir T., 152
degree of infringement, 179, 182, 185, 186, 195
*Derby Mercury*, 37, 52, 214
Derosne, Charles, 74, 158
Dickens, Charles, 51, 58
Dickinson, H. W., 10, 52, 169, 222
diffusion, definition, 11
Dircks, Henry, 24, 31, 87, 94, 99, 119, 123, 142, 143, 145, 149, 170, 187, 199, 215
disclosure, 75–6
Dod, J., 87
Dollands case, 71, 188
Donaldson, John, 135, 136
Donkin, Bryan, 156
Donkin, H. I., 98, 215
Dossie, Richard, 157
Dove, George Godney, 184
Downton, J., 158, 162
drainage, 165–8
Drewry, C. S., 119, 215
Dundonald, Lord, 26, 32, 133, 137–9, 141, 154
Duncan, John, 152, 181
Dundas, Lord, 137–9
Dundee, 124
Durnford's, 127, 129
Dutton, H. I., 120, 199, 224

Earnshaw, Thomas, 212
Eaton, Mr., 90
economic determinism, 107–8, 142, 203
*Economist*, 23, 25, 29, 31, 32, 68, 172, 214
Ede, J. F., 148, 197, 218
Edelston and Williams, 126
Edwards, F., 25, 28, 32, 119, 215
Eldon, Lord, 22, 74, 77
electrical industry, 131
Ellenborough, Lord, 78, 199
Elsas, M., 199, 218
Else, A., 37–8, 52
Encel, S., 119, 222
*Engineers' Encyclopaedia*, 20
Enrolment office, 87, 198
Erskine, Mr., 41
exchange-for-secrets thesis, 17, 22; see disclosure
externalities, 175
Eyre, Justice, 71, 73, 79, 82

Failsworth, 90
Fairnbairn, William, 58, 60, 63, 90, 160, 184
Faire, John, 29, 83
Farey, John, 21, 22, 31, 35, 44, 45, 47, 49, 50, 60, 67, 70, 79, 82, 88–9, 93, 98, 135, 182, 187, 193, 194
Farrer, T. H., 32, 215
Fay, C. R., 65, 68
Fearon, P., 196, 218
Felkin, William, 84, 97, 146, 199, 201, 215
Ferrabee, John, 160
Field, Joshua, 29
Finer, S. E., 68
Finzel, Mr., 135, 136
Fischer, J. C., 21, 136
Fisher, Mr., 192

Fitton, R. S., 171, 172, 199, 219
Fitzmaurice, Lord, 66, 215
flax industry, 126, 134–5, 136, 158, 161, 181–2, 185, 189–92
Ford Bros. and Co., 163
Forwell, W. R., 97, 219
Fothergill, Benjamin, 61, 95
Fourdrinier, Henry, 156, 161, 213
Fourdrinier, Sealy, 156, 161, 213
Fowler, Thomas, 95
Fox, H. G., 3, 11, 219
Freeman, C., 170, 219
Frumkin, M., 11, 222
Fry Committee, 68
fustian tax, 36, 37

Gamble, John, 156, 161
Gardiner, Mr., 162
Gardum, Pare and Co., 137
gas industry, 44, 154, 182
Gattrell, V. A. C., 195, 222
Gauldie, E., 143, 219
Gedge, J., 87
Gellson, Mr., 130
General Chamber of Manufacturers, 36
Gilbert, Mr., 41, 43
Gilfillan, S. C., 105, 118, 219, 222
Gill, Thomas, 87, 90, 215
Glasgow, 29, 36, 60
glass industry, 126, 131, 133–4, 135, 137–8
Gloucester, 126, 127
Glyn, George, 167
Godson, Richard, 30, 46, 47–8, 50, 55, 82, 83, 84, 173, 196, 215
Gomme, A. A., 1, 10, 51, 83, 222
Gordon, J. W., 215
Gosset, F. R. M., 161
Grace, Stanford and Berridge, 194
Grant, Thomas, 58
Granville, Lord, 25, 28, 58–9, 60, 62
Great Exhibition, 21, 58, 79
Greathead, Henry, 212
Greathead, P., 165–8
Great Seal office, 35, 36
Greenock, 29
Greville, Mr., 87
Grey, Earl, 166
Grimshaw, E., 190
grinding corn, 193
Guest, Josiah, 185
gunmaking, 156
Gurney, Goldsworth, 162, 169, 213

Habakkuk, H. J., 8, 153–4, 170, 195
Hackwood, F. W., 197, 215
Haddan, J. C., 87
Hales, W. S., 131
Halifax, 59, 63
Hall, English and Co., 127, 129
Hall Horace, 189
Hall, Mr., 88
Hamilton, W., 11, 222
Hamilton, William, 139
Hammond, J. L. and B., 85
Hancock, Charles, 131, 141
Hancock, T., 161, 172, 215
Handcock, E. R., 132
Handley, Mr., 184
Hands, William, 74, 83, 215
Hanning, Mr., 162

# Index

Hard, Mr., 87
Hardie, D. W. F., 148, 221
Harding, A., 81, 219
Harding, H., 1, 3, 10, 11, 45, 54, 68
Hardy, J., 126
Hargreaves, James, 188
Harlow, V. T., 52

Harris, Timothy, 127, 144
Hart, J., 68
Hartley, David, 33, 79
Hartley, James, 133–4
Hartwell, R. M., 11, 97, 100, 219
Hately, Joseph, 185
Hatfield, S., 3, 11, 219
Hawden, John, 57
Hayek, F. A., 81
Hayes, John, 136
Hayward, E., 162
Heath, Joseph, 193
Heath, Justice, 74
Heathcote, John, 27, 49–50, 77, 88, 134, 135, 136, 185, 193, 194
Heilman, J., 126, 135
Henderson, W. O., 31, 198, 219
Henson, Gravenor, 27, 42, 51, 52, 136, 181, 182
Herbert, Luke, 20, 31, 87, 90
Hill, Anthony, 185
Hill, Mr., 162
Hill, Samuel, 212
Hindmarch, William, 18, 30, 31, 51, 75, 76, 81, 83, 84, 99, 146, 164, 170, 173, 201, 215
Hives, Mr., 192
Hobson Smith, J., 182
Hodges, Paul, 90, 95
Hodgkinson, E., 90
Holden, Isaac, 160, 162
Holderness, B. A., 174
Holdsworth, A., 201
Holdsworth, W., 3, 11, 68, 219
Holland Rose, J., 52
Hollingrake, Mr., 132
Hollins, Samuel, 165
Holroyd, E., 143, 169, 215
Hooper, F. D., 181
Hopkins, Roger, 132
Horrockses, Miller and Co., 156
Howard, Mr., 135, 136
Howe, J. I., 130
Howe Manufacturing Co., 130
Hume, Joseph, 42, 64, 111
Hunt, James, 166–7, 185
Hunt, Ogle, 166–7
Hunt, Seth, 128
Hunt, William, 129
Hyde, C. K., 117

industrial espionage, 47, 111, 119, 181
industry, problems of definition, 114–5
Ingleby, Rupert, 132
Inglis, A., 119, 222
innovation, definition, 11
Institute of Civil Engineers, 88
Institute of Mechanical Engineers, 122
institutional arrangements, 175
invention
    and science, 105, 117, 118
    definition, 11
    demand for, 105–6, 108, 203
    expectations, 107–8
    industry, criteria for, 108, 123–4
    market for, 124–5, 203
    supply of, 107–8
    trade in, 108, 122–42, 203
inventive activity, 103–17
    bankruptcy, 106–7
    direction of, 113–6, 203
    diversification of, 108, 113–6, 119, 142
    economic growth, 103
    investment in, 93, 150–69
    outside the user industry, 122–4
inventors
    as captains of industry, 107–8
    captive, 106
    heroic, 105, 113, 117
    motives of, 104–6, 108–9
    multiple, 112–6, 120, 142, 159, 168, 203
    quasi-professional, 108, 112–6, 124, 125, 128
Inventors' Aid Association, 59, 60, 63
Inventors' Law Reform League, 59
Ireland, 35, 36, 61, 63
Irish Commercial Treaty, 36, 37, 39, 188
iron industry, 44, 90, 122, 132, 135, 136, 156, 176, 184, 193

James, Francis, 188
James, J., 103, 117, 216
Jenkinson, William, 133
Jeremy, D. J., 143, 219
Jewkes, J., 8, 12, 105, 110, 118, 143
Johnson, H. G., 11, 222
Johnson, John Henry, 97
Johnson, P. S., 118, 219
joint-stock companies, 163
Jones, E. L., 53
Jones, S. R. H., 143, 199, 222
Jones, T., 157
Jones, William, 129–30
*Journal of the Society of Arts*, 33, 67, 68, 196, 214
*Journal of the Statistical Society of London*, 214
Judical Committee of the Privy Council, 48, 50, 57, 65, 125, 151, 154, 155, 159, 171, 191

Kaempffret, W. B., 118, 219
Kammen, M., 52
Kay, Alexander, 47
Kay, James, 134, 136, 158, 181, 189–92, 200
Keir, James, 183
Kendrick, W., 20, 31, 216
Kennedy, C., 118, 222
Kennedy, J., 162
Kenyon, Lord, 21, 31, 41, 72, 74, 77, 84, 217
Kieve, J., 145
Kingston, W., 11, 219
Kindleberger, C. P., 118
King, J. E., 199
Koch, J. V., 12
Koebner, R., 142
Koop, Matthias, 163, 212
Kuznets, S., 13
Kyan, John Howard, 135, 137, 163, 213

Labouchere, H., 58
lace industry, 88, 134, 185, 192, 193–4
Lancaster, 60
Landes, D. S., 3, 11, 103, 117, 150, 151, 154, 169, 220
Lane, R. K., 161, 173
Langdale, Lord, 198

# Index

Langford, J. A., 52, 216
Langton, J. K., 169, 216
Langton, Stephen, 95, 141, 163, 178, 183
Law, Sam, 190
Law Society, 94
lawnmowers, 160
Lea, James, 58
Leblanc process, 139
Lee, James, 42, 212
Ledsam, Daniel, 129–30, 182
Leeds, 46, 181
Leek, 60
Lefevre, Mr., 58
Lehmberg, S. E., 51
Leith, 29
Lennard, T., 20, 22, 43–4, 46, 96
Lewis, John, 192
licensing, 93, 124, 132–40, 141, 146, 152, 159, 165, 167, 168, 194
Lilley, S., 12, 103, 117, 220
limited liability, 152–3, 163, 166
Lister, Samuel, 135, 160
Liverpool, 29, 59, 60, 184
Liverpool Coal and Gas Co., 140
Liverpool, Lord, 39
Llanelly Copper Co., 132
Lloyd, J. H., 29
Lloyd, Samuel, 132
London, 36, 42, 49, 51, 59, 60, 63, 87, 88, 90, 128, 130, 136, 140, 168
Losh, William, 138, 141
Loughborough, Lord, 80, 109
Lyndhurst, Lord, 55, 97, 167

MacCleod, H. D., 30, 216
MacDonagh, O., 65, 68
MacDonald, Sir, Archibald, 40–1
MacFie, R. A., 24, 28, 29, 81, 135, 143, 146, 185, 197, 216
MacKinnon, W. A., 49, 50, 51, 57, 217
Maclaurin, W. R., 12, 222
MacLeod, C., 30, 51, 83, 120, 143, 170, 224
machine-making, 44, 90, 133, 160, 185
Machlup, Fritz, 5, 12, 30, 205, 220, 222
Macintosh, Charles, 141, 161, 216
Macintosh, G., 172, 216
Manchester, 36, 37, 43, 44, 46, 47, 59, 63, 87, 88, 90, 129, 130, 140, 161, 162, 188
Manchester Chamber of Commerce, 42, 60
Manchester Commercial Committee, 27
*Manchester Guardian*, 66, 67, 68, 90, 98, 214
Manchester Law Society, 47
*Manchester Mercury*, 27, 32, 188, 199, 214
Manchester Patent Reform Committee, 58, 60, 62–3, 64, 68
Mann, Julia de Lacy, 200, 220
Mansfield, E. A., 12, 118, 220
Mansfield, Lord, 22, 71, 75, 81
March, James, 49
Mardon and Co., 125
marine patents, 115, 158, 160, 162
Marling, N. S., 192
Marlowe, Allicott and Seyrig, 132
Marsh, Mr., 37
Marshall, John, 182, 185, 189–92
Marshall and Co., 126, 134, 140, 161, 189–92
Marshall, L. S., 199
Martineau, J., 136, 182
Martineau, P., 160
masters' associations, 187

Mathias, P., 169
Matthews, Mr., 38, 40–1
Matthews, R. C. O., 145, 173
Maudsley, Henry, 29
Maurice, Mr., 37
May, Charles, 60, 181
M'Kay, Dr., 162
McCloskey, D. N., 203, 205, 220
McCulloch, J. R., 17–8, 30, 45, 54, 216
McKendrick, N., 148, 149, 169, 222
McKie, D., 10
*Mechanics' Magazine*, 22, 29, 30, 31, 33, 43, 44, 46, 47, 49, 50, 51, 54, 55, 56, 59, 62, 64, 66, 67, 68, 79, 84, 87, 96, 100, 112, 119, 120, 149, 172, 214
Meikle, Andrew, 193
Mercer, John, 163
Merry, Joseph, 44, 70
Messrs Greenfell, 160
Messrs Kendal and Alton, 185, 193
Messrs Ley and Mason, 191
Messrs Nunn Brown and Freeman, 77, 185, 193
Messrs Sharp, 162
Messrs Sheerman and Evans, 183
Messrs Terry and Parker, 135, 136
Meteyard, E., 199, 216
Mill, J. S., 19–20, 30, 58, 216
Mingay, G. E., 53
Mishan, E. J., 195
Monmouthshire Iron and Coal Co., 132
monopoly-profit thesis, 17, 20–2
Mordaunt, Col., 188
Morgan, William, 158
Morland, James, 139–40
Morley, Williams, 185
Morton, S., 28, 158, 159, 212
Morton, T., 179, 181
motive power, 115, 116, 158
Motley, Thomas, 58
Moulton, Samuel, 93–4
Muirhead, J. P., 82, 216
Mulholland, J. K., 190
Mulholland, Mr., 190, 191
Muntz, G. F., 154, 160, 182
Murdock, William, 178
Murray, Matthew, 141, 161, 183
Mushet, David, 126, 132, 184
Mushet, William, 126
Musson, A. E., 10, 32, 53, 83, 97, 107, 117, 118, 119, 147, 148, 172, 195, 197, 198, 220

nailmaking, 156
Nasmith, James, 75
Nasmyth, J., 7, 90, 179
National Association for Reform of Patent Laws, 57
National Patent Law Reform Association, 58, 59, 63
natural-law thesis, 17–8
Neale, J. E., 10
Neale and Palmer, 194
Need, Samuel, 160
needle making, 156
Neilson, J. B., 135, 136, 182, 193
Nelson, R. R., 13, 118, 145, 149, 222–3
Netherlands, 9, 11, 29
Nevill, E., 132
Newcastle, 137
Newcastle and Gateshead Chamber of Commerce, 94–5
Newton, A. V., 87, 90, 91, 95, 98, 131, 216

# 230   Index

Newton, William, 45, 50, 55, 60, 67, 77, 86, 87, 90, 95, 96
Newton, William Jnr., 90, 97, 216
*Newtons Journal*, 54, 55, 66, 67, 83, 96, 172, 200, 214
Neymeyer, F., 118
Nightingale, Peter, 189
Nordhous, W. D., 11, 220, 223
Norman, J. P., 169, 216
Norris, J. M., 52, 223
Norris, K., 118, 220
North, D. C., 175, 195, 220
Norwich, 36
Nottingham, 37, 132, 140, 188, 189

O'Brien, D. P., 30
Ogburn, W. F., 105, 118
Old Park Iron Works, 132
Olson, M., 30
optical instruments, 188

paper industry, 136, 156–7, 161, 163
Parke, Baron, 78, 80
Parker, J. E. S., 13, 220
Parker, Mr., 87
Parker, Mr., 154
Parks, Mr., 50
parliamentary awards, 25–6
Parris, H., 68
partnerships, 124, 140–1, 148, 160–3, 169
Pasman, John, 157
patents
    administration, 34, 35, 59, 63
    and economic growth, 2–3, 10, 20–2, 68, 204–5
    caveats, 35, 95, 111, 183–4
    costs, 34, 35, 42, 44, 45, 48, 50–1, 59, 60, 61–2, 63, 110–11, 203
    efficiency of patent protection, 26–7, 34, 36, 44, 59
    extensions, 28, 48, 50, 155
    fluctuations in patent activity, 176–8, 204
    free trade, 24
    market power, 7, 9
    monopolies, 22–3
    number of, 3, 176
    pay-off period, 44, 151–2, 159
    profits, 108–9, 142, 159
    rate of profit, 153–4
    rewards, 25–6
    specification, 22, 36, 39–40, 42, 44, 75–6, 90, 92, 93, 184–6, 198, 199
    statistics, problems of, 1, 6–7, 110–12, 203
    survival rate, 159, 172, 209
Patent agents, 86–96, 183, 184
    criticisms of, 94–6
    functions, 91–4
    patent officials, 86–8, 96
    provincial agents, 88
    training, 90–1
*Patent Journal*, 32, 54, 64, 67, 68, 79, 84, 120, 178, 196, 214
Patent Law Amendment Act, 1, 17, 34, 57–65
patent law, 69–81, 194–5
    definition of patent, 47–8, 72–5
    effect on patenting, 176–8, 195
    effect on value of patents, 72, 195
    improved nature of, 78–81, 175, 180
    number of cases, 70–2, 125, 180
    uncertainty of, 34, 44, 69–70, 176, 180, 195, 203
    unofficial testing of patents, 72, 195
Patent Law League, 58, 60
Patent Office, 6, 7, 26, 35, 36, 44, 45, 48, 60–1, 69, 87, 88, 90, 92, 96, 111, 167, 182, 183, 184, 185, 198
Patent Office officials, 44, 45, 48, 62, 86, 87, 111
patent reform, 34–67, 203–4
    before 1820, 36–42
    in the 1820s and 1830s, 42–51
    in the 1840s and 1850s, 57–65
    influence of Lord Brougham, 48, 51, 65
    provisions of 1852 Act, 63
    reactions to 1852 Act, 63–4
    slowness of reform, 64–5
    Watt's proposals, 39–41
Patent Reform Club, 59
Patentees' Association, 37, 38, 40, 70
Paul, Lewis, 157
Payne, Mr., 87
Payne, P. L., 99, 145, 173, 220
Peacock, J., 216
Peel, Robert, 37, 44, 45, 189
pencil making, 125
Penn, Mr., 29
Penrose, E., 30
Perkins, A. M., 154, 157
Petty Bag office, 45, 198
Phillpots, Mr., 46
Phipps, J., 136
Pigou, A., 105, 118
Pilkington, Mr., 134
pin making, 126–30
Pitt, William, 36, 37, 39, 188
Place, Francis, 42, 64
Plant, A., 4, 11, 223
Pole, W., 171, 220
Political Economy Club, 30–1
Pollard, S., 169, 171
Poole, James, 86
Poole, Moses, 45, 60, 87, 88, 95, 96, 97, 184
Porter, Mr., 157
Post, R. C., 111, 120, 223
pottery industry, 194
Prager, J., 83
Pratten, C. F., 144
Pressnell, L. S., 170
Preston, 156
price associations, 187
Price, W. H., 10, 31, 219
Price's Candle Co., 131
Prime, Mr., 50
Prince, Alexander, 31, 87, 214, 216
Prince, J., 95
Prior, Thomas, 58
Privy Seal office, 35
property rights, 175, 176
Prosi, G., 11, 12, 220, 223
Prosser, R. B., 88, 95, 97, 98, 144, 145, 159, 216
public utilities, 163
Pugh, Evan, 138
Pultney, Sir William, 21, 31

*Quarterly Review*, 81, 214
Queen Elizabeth I, 1

Radnor, Lord, 41
Rae, J., 216
railways, 126, 132, 183, 185
Raistrick, John, 193
Ralph and Spooner, 140

# Index 231

rate of interest, 153, 170
Ravenshear, A. F., 3, 11, 216
Raymond, M. Amedée, 129
Reade, Compton, 161
Record and Commissioners office, 63
Redford, A., 32
Reed and Aberlady, 165
*Rees Encyclopaedia*, 88, 144, 214
Reeve, Henry, 150, 169
Renshaw, William, 189
*Repertory of Arts*, 21, 67, 151, 197, 199, 214
residual, 103, 117
Retrie, John, 154
revolution in government, 65
reward-by-monopoly thesis, 17, 18–20
ribbon manufacture, 44
Ricardo, J. L., 25, 29, 111, 131
Richardson, Joshua, 185
Riddle and Piper, 157
Rimmer, W. G., 143, 172, 190, 199, 200, 220
Robbins, Lord, 30
Roberts, D., 65, 68
Roberts, Richard, 46, 47, 54, 60, 110–11, 131, 132, 141, 157, 159, 160, 216
Robertson, Joseph Clinton, 43, 44, 47, 87, 90, 93, 96, 98, 216
Robinson, Eric, 2, 10, 13, 32, 53, 81, 82, 83, 97, 117, 119, 147, 148, 172, 197, 198, 220, 223
Robinson, Joan, 3, 11
Robinson, Solomon, 135–6, 200
Robson, R., 97
Rochdale, 59, 63
Rogers, J. E. T., 18, 30, 216
Roll, E., 53, 169, 221
Rolls Chapel office, 87, 198
Romilly, Lord, 29
Rosenberg, N., 4–5, 9, 11, 13, 30, 105–6, 118, 119, 142, 221, 223
Rosser, Archibald, 46, 47–8, 49, 50, 61, 83, 100, 148, 158
Rossman, J., 118, 221
Rostow, W. W., 103, 117, 195, 196
Rotch, Benjamin, 43, 46, 47–8, 87, 95
Royal Society, 40
rubber industry, 94, 131, 132, 161
Russell, James, 74, 136, 157, 162, 181, 182
Russell, John Scott, 29, 193
Ruttan, V. W., 13, 223

Scherer, F. M., 11, 12, 119, 223
Scherer, H., 172, 223
Schiff, E., 5–6, 12, 221
Schmookler, J., 7, 9, 12, 15, 105–6, 118, 119, 120, 221, 223
Schofield, R. E., 147, 221
Schomberg, C. L., 126
Schumpeter, Joseph, 7, 104, 118, 221
Scoffern, Dr., 135
Scotland, 35, 61, 63, 82, 165, 188
*Scotsman*, 30, 214
Seaborne Davies, D., 83
search costs, 93
search process, 130–2
seasoning wood, 163, 178
secrecy, 6, 42, 110–11, 204
selling inventions, 47, 93, 120, 125–32, 143, 159
Shaftsbury, Lord, 168
Sharpe, Mr., 141
Sharpe, W. F., 116, 121
Shaw, Major, 43

shearing machines, 192
Sheffield, 36
Sheppard, Sir, Samuel, 97
Sherbourne, Robert, 138
Shrapnel, Henry, 58
Shrewsbury, 161
Signet office, 35
Sigsworth, E. M., 172
Silbertson, Z. A., 1, 3, 10, 11, 26, 32, 51, 145, 201, 224
silk manufacturing, 90
Simpson, W. H., 141
Slocum, Samuel, 130
Small, Dr., 73
Smiles, Samuel, 1, 11
Smith, Adam, 18–9, 30, 106–7, 119, 122, 124, 126, 142, 174, 195, 216
Smith, E. M., 157
Smith, J. W., 216
Sneath, William, 192
social and private rates of return, 175
socially wholesome illusion effect, 205
Society for the Amendment of Patent Law, 60
Society of Arts, 25, 58, 62, 64, 111, 157
Society of Inventors, 51, 59, 63
Society for Promoting Scientific Inquiry, 58
Solo, C. S., 13, 224
Southworth, William, 158
Spackman, Mr., 188
*Spectator*, 23, 31, 214
Spence, William, 23, 60, 87, 90, 216
Spencer, George, 131, 161, 173
Spencer-Moulton, 131
Stafford, Daniel, 154, 158
Stamp, Sir J., 105, 118, 221
Standards of patentability, 111
Stanhope, Lord, 41
Stark, W., 30, 68, 221
Statute of Monopolies, 17, 28, 47, 50, 72, 73, 80
Staveley, C., 88
Steam, 74, 88, 109, 122, 158, 176, 178
steam carriages, 162
steel, 184
Stephens, G., 165–6
Stephenson, Robert, 166–7
Stewart Smith and Co., 182
Stockport, 90
Strassman, W. P., 13, 221
Stroud, 129, 192
Strutt, Jedediah, 140, 160, 188
Stubs Wood, 128, 129
sugar refining, 132, 135, 136, 158, 160
Surrey, James, 184
Surtees, Aubone, 137–9
Sutherland, G., 68
Swansea, 183
Switzerland, 5–6, 29
system of registration, 111

Taitt, William, 185
Tann, J., 149, 224
Taussig, F. W., 118, 217
Tayler, D. F., 129
Tayler, Shuttleworth and Watnerby, 129
Taylor, C. T., 1, 10, 11, 145, 201, 224
Taylor, John, 44, 45, 70, 197
Taylor, O. H., 30
Taylor, Samuel, 52
Taylor, William, 188
Tayman, F., 140

# 232  Index

technological change, 8–9, 202
Tenterden, Lord, 78
textiles, 115, 116–7, 157–8, 176, 188, 192
*The Engineer*, 214
*The Times*, 23, 31, 43, 54, 96, 98, 100, 197, 198, 214
Thirwall, A. P., 118, 222
Thisselton, Mr., 162
Thomas, Edward, 183
Thomas, R. P., 175, 195, 220
Thompson, F. M. L., 97
Thompson, R., 70
threshing machine, 193
tiles, 140, 141, 164–8, 185
Tiverton, 185
Tooke, M., 50
total factor productivity, 202
Toynbee, A., 103, 117
trade cycle, 153, 177–8, 204
*Transactions of the Institute of Patent Agents*, 97, 98, 99, 214
*Transactions of the Society of Arts*, 32, 120, 214
tube industry, 136, 140, 159, 162
Turner, I., 133, 179
Turner, T., 84, 217
Tweeddale, Marquess, 140, 141, 164–8, 185

United Inventors' Association, 58, 60
Uselding, P., 2, 11
Usher, A. P., 8–9, 13, 221

Vaizey, J., 118, 220
Vallet, Matthew, 28
Vaughan, F., 118, 221
ventilators, 125
Viney, Sir, James, 162
von Tunzelman, G. N., 117, 205, 221

Wadsworth, A. P., 171, 172, 199, 219
Wagner, D. O., 31
Wales, 132
Walker, B. W., 191
Walshamn, T., 137
Warden, A. J., 199, 217

Warrington, 126, 128
Watson, Mr., 41
Watt, James, 26, 36, 37, 38, 39, 40, 41, 73, 74, 76, 82, 86, 109, 136, 140, 151, 153, 158, 159, 180, 182, 183, 184
Watt, Junior, 74, 178
Weaver, Charles, 128
Webb, R. K., 85
Webster, Thomas, 18, 23, 30, 31, 32, 43, 45, 54, 57, 59, 60, 61, 62, 65, 67, 68, 81, 82, 83, 84, 85, 97, 124, 132, 144, 149, 155, 159, 170, 171, 172, 173, 182, 197, 200, 217
Wedgwood, Josiah, 26–7, 32, 39, 40, 83, 184, 194
Wednesbury, 132
West, E. G., 65, 68
Westhead, J. P., 60
*Westminster Review*, 18, 30, 70, 73, 74, 81, 82, 83, 84, 214
Weston, Abraham, 38–9, 41, 70, 183
Wheatley, Mr., 87
Wheatstone, C., 131
Wheeler, J., 52, 217
Whitehouse, James, 140, 157, 162
white lead manufacture, 133
Whitworth, Joseph 90
Whytehead, William, 58
Williams, B. R., 118, 218
Winter, S. G., 145, 149, 222–3
Wise, Lloyd, 94, 99, 217
Wittacker, Mr., 185
Wood, Benjamin, 138–9
Wood, T. H., 32, 217
Wood, William, 138–9
Woodcroft, Bennet, 61, 81–2, 87, 90, 98, 112, 115, 119, 120, 144, 158, 217
Woodruff, W., 145, 172, 221
wool combing, 160
woollen industry, 123, 135, 141, 193
worsted manufacture, 162
Wright, Lemuel, 128–9, 159
Wrottersley, Mr., 41
Wyatt, John, 157
Wyatt, W. H., 21, 31, 67, 151, 199, 217
Wyndam Hulme, E., 81, 83, 217